To Pat

With compliment

Lo

History of Mechanism and Machine Science

Volume 29

Series Editor

Marco Ceccarelli
Cassino, Italy

Aims and Scope of the Series

This book series aims to establish a well defined forum for Monographs and Proceedings on the History of Mechanism and Machine Science (MMS). The series publishes works that give an overview of the historical developments, from the earliest times up to and including the recent past, of MMS in all its technical aspects.

This technical approach is an essential characteristic of the series. By discussing technical details and formulations and even reformulating those in terms of modern formalisms the possibility is created not only to track the historical technical developments but also to use past experiences in technical teaching and research today. In order to do so, the emphasis must be on technical aspects rather than a purely historical focus, although the latter has its place too.

Furthermore, the series will consider the republication of out-of-print older works with English translation and comments.

The book series is intended to collect technical views on historical developments of the broad field of MMS in a unique frame that can be seen in its totality as an Encyclopaedia of the History of MMS but with the additional purpose of archiving and teaching the History of MMS. Therefore the book series is intended not only for researchers of the History of Engineering but also for professionals and students who are interested in obtaining a clear perspective of the past for their future technical works. The books will be written in general by engineers but not only for engineers.

Prospective authors and editors can contact the series editor, Professor M. Ceccarelli, about future publications within the series at:

LARM: Laboratory of Robotics and Mechatronics
DiMSAT—University of Cassino
Via Di Biasio 43, 03043 Cassino (Fr)
Italy
Email: ceccarelli@unicas.it

More information about this series at http://www.springer.com/series/7481

Leslie Philip Pook

British Domestic Synchronous Clocks 1930–1980

The Rise and Fall of a Technology

 Springer

Leslie Philip Pook
Sevenoaks, UK

ISSN 1875-3442   ISSN 1875-3426 (electronic)
History of Mechanism and Machine Science
ISBN 978-3-319-14387-3   ISBN 978-3-319-14388-0 (eBook)
DOI 10.1007/978-3-319-14388-0

Library of Congress Control Number: 2014958589

Springer Cham Heidelberg New York Dordrecht London

Springer International Publishing AG Switzerland is part of Springer Science+Business Media (www.
springer.com)

Preface

The first clock I owned was an Ebosa 30-hour alarm clock, a 16th Birthday present from my parents. This woke me up reliably until several years later when it stopped working. By this time I had rebuilt two motorcycle engines, so it seemed to me that I ought to be able to repair the Ebosa. I took it apart and found that a tooth had broken off one of the pinions. I took the pinion, still on its arbor together with a wheel, to Clerkenwell, the clock making area of London, where I found a friendly dealer who sold me a replacement. After re-assembly the clock worked and was used for several more years until wear made it unreliable.

Living with my parents, and later in lodgings, did not leave much scope for collecting, but my experience with the Ebosa, and some other balance wheel clocks, had convinced me that collecting clocks was practical since I could do my own clock servicing and repairs. Soon after we were married in 1960, my wife and I bought an eighteenth-century long-case clock by John Barrow, London, for £9 10/-. My confidence in my clock making abilities was not misplaced, and the clock is still running and keeping good time. Encouraged by my wife further pendulum and balance wheel clocks followed.

I had seen a Selectra synchronous electric mantel clock, which I liked, at a relative's house. Coincidentally, an identical clock turned up at the Summer Fayre held at my son's school in 1983. I bought the clock for 50 pence. It was not in working order, but I managed to get it going. Further domestic synchronous clocks followed. Prices have gone up, but interesting examples can still be obtained for modest sums. Current electrical regulations mean that professional clockmakers are reluctant to repair synchronous clocks. In fact, provided that they have not been mistreated, synchronous clocks are usually reliable, and quite easy to maintain. Cleaning and oiling is often all that is needed to get a reluctant clock to run. Synchronous movements, probably removed from clocks that have been converted to quartz, are sometimes listed on eBay. This is a useful source of spares that are otherwise unobtainable. The advent of digital photography has made servicing an unfamiliar synchronous clock, for which no instructions are available, much easier since photographs can be taken during dismantling as a guide to subsequent re-assembly.

Six decades after acquiring my first clock my hobby has progressed to a point where I now have an extensive archive of information on domestic synchronous clocks, their movements, and their manufacturers. Some of this is from published sources, but much of it has been derived from examination and photography of clocks. I set up my website *The Les Pook Miscellany*, www.lespook.com, in 2009. This is updated regularly as new information becomes available. There has been a lot of traffic to the synchronous clock pages, and many useful comments from other enthusiasts. Other topics covered include metal fatigue, fracture mechanics, Meccano, and flexagons.

I have extensive experience of publishing in my professional areas of metal fatigue and fracture mechanics. I decided to publish some of my domestic synchronous clock archive, and in January 2014 my article *Temco Art Deco Domestic Synchronous Clocks* appeared in *Watch & Clock Bulletin*. The article was well received, and I decided to write this book on British domestic synchronous clocks. It is primarily for fellow enthusiasts, and complements available one-make books on domestic synchronous clocks. It is also a history of science book that sets British domestic synchronous clock technology in its social context.

The profusely illustrated book is in two parts. Part I covers the historical background, British domestic synchronous clock manufacturers and brands, how synchronous clocks work, domestic synchronous clock cases, practical advice on the servicing of domestic synchronous clocks, and analysis of the marketing and reliability of British domestic synchronous clocks. This analysis provides an explanation of the rise and eventual fall of their technology. Part II contains galleries of a selection of British domestic synchronous clocks, and of the movements with which they are fitted. There is a front and back view of each clock, together with a brief description. Views of each movement include views with the movement partly dismantled, together with a brief technical description of the movement. The detailed Contents serves as an index.

Thanks for assistance are due to staff at the British Library and at the British Standards Institution. Thanks are also due to Internet contacts who have provided information. Where appropriate they are acknowledged in the text.

I retired formally from University College London in 1998 but remained active in my profession fields of metal fatigue and fracture mechanics. Retirement left me with more time to pursue long standing interests in recreational mathematics and horology. My association with Springer started in 2005. This is my fourth book published by Springer. The previous books are *Metal Fatigue*, *Serious Fun with Flexagons* and *Understanding Pendulums*.

November 2014   Leslie Philip Pook
Sevenoaks, UK

Contents

Glossary and Abbreviations

AC	Alternating current
ADD	Anti-dumping duty
Alarm clock	An alarm clock is a bedside clock with a time switch and an audible alarm as complications
Arbor	Arbor is the horological term for a shaft or axle
Atomic clock	An atomic clock is based on structural transitions which takes place within an individual atom. Atomic clocks are astonishingly accurate. The first atomic clock was built in 1971. Atomic clocks are not used as domestic clocks
Back set	A back set movement has the hand set knob at the back of the clock
Bedside clock	A bedside clock is a clock intended for display on a bedside table, but can also be used as a desk clock. Some bedside clocks have a time switch as a complication
Bottom set	A bottom set movement has the hand set knob at the bottom of the clock
Case	The case is the externally visible part of a clock, usually including the dial and hands, but excluding any visible parts of the movement. It encloses the movement to protect it
Clock	In modern usage, clock refers to any device whose primary function is measuring and displaying time, but not intended for carrying on the person. Here, it is restricted to a device with an analogue display
Coercivity	Coercivity is the extent to which a material retains magnetism when it is removed from a magnetic field

Complication	A complication is a secondary function of a clock. Complications include striking and chiming, which provide an audible indication of the time, and time switches
Counter train	A counter train is a train of gears that counts and sums pulses that measure time into hours and minutes, etc.
Crossed out	A wheel that is crossed out has material removed to leave a rim connected by spokes to a central hub. This is done to reduce the inertia of wheels that move intermittently, as in mechanical clocks
DC	Direct current
DIY	Do it yourself
Dumping	Dumping is the sale of goods at a price considerably less than their normal value. This is defined as a price that is lower than the price of similar goods in the country from which they originate
Electric clock	An electric clock is a clock whose power source is electricity. These include synchronous clocks, electrically driven pendulum electric clocks, and quartz clocks
Electromagnet	In an electromagnet a temporary magnet is produced by a coil through which an electric current is passed
Escapement	An escapement is a controlling device which converts oscillations into a series of pulses to measure time. It also provides impulses to the oscillator to replace energy loss
Going train	A going train is the counter train of a mechanical clock
Granddaughter clock	A granddaughter clock is a longcase clock that is less than 157 cm tall. It has a dial that can be read from across the room
Hybrid pinion	This is a coined term. Horologically, a hybrid pinion is a lantern pinion with only two rods
Indicating device	An indicating device is usually a dial with hands that indicate hours, minutes and sometimes seconds
Insertion movement	An insertion movement is a movement enclosed in a movement cover and fitted with a dial and hands. It can be inserted as a unit into an appropriate case
Magnet	A magnet is a material or object that produces a magnetic field

Magnetic field	A magnetic field is the invisible field responsible for the most notable property of a magnet: a magnetic force that attracts or repels other magnets
Magnetic flux	Magnetic flux is the strength of a magnetic field at any point
Magnetic reluctance	The magnetic reluctance of a material is a measure of its resistance to a magnetic field. It is analogous to a material's electrical resistance
Mainspring	A mainspring is a coiled spring used as the power source for a mechanical clock
Mantel clock	A mantel clock is a clock intended for display on a mantelpiece or on a piece of furniture. It has a dial that can be read from across the room
Marketing	For consumer goods, such as domestic clocks, marketing has two different meanings. The context will usually show which is intended. (1) Marketing can be regarded as an organisational function and a set of processes for creating and distributing goods to consumers, and improving the user experience. Marketing is also the science of choosing target markets through market analysis, as well as understanding consumer behaviour, and providing superior customer value. (2) Marketing is the process of distributing goods to consumers
Marriage	A marriage is a clock in which the original movement has been replaced by a movement of a different type
Master clock	A master clock is a high-precision pendulum electric clock which, as a complication, provides electrical impulses to drive subsidiary dials. For scientific purposes, master clocks were superseded by quartz clocks
Mechanical clock	A mechanical clock is a clock whose power source is mechanical. Mechanical clocks for domestic use are still available, but have been largely superseded by quartz clocks
Mechanical time switch	A mechanical time switch has a mechanical movement similar to the movements used in mechanical clocks. The switches are mechanical. They were superseded by synchronous time switches
Movement	The movement is the mechanism of a clock which drives the hands and any complications

Novelty clock	A novelty clock is a clock that is primarily an ornament. In general, synchronous novelty clocks are not suitable for use as domestic clocks
Outage indicator	An outage indicator is a device on a self-starting synchronous clock which shows that there has been a power outage so the clock is showing an incorrect time
Permanent magnet	A permanent magnet is made from a magnetically hard material placed within a magnetic field. It has its own persistent magnetic field
Power source	A power source replaces energy lost due to inevitable friction etc. In a mechanical clock the power source is usually either a falling weight or a mainspring. In an electric clock the power source is either a battery or AC mains
Quartz clock	A quartz clock has a quartz crystal as a high frequency oscillator. The power source is usually a battery providing low voltage DC. The quartz crystal communicates with the power source through the piezoelectric effect. An electronic frequency divider provides the equivalent of a reduction gear. Quartz clocks for retail sale were first produced in 1971. They are sometimes called electric clocks. Quartz clocks are sometimes subsidiary dials controlled by radio signals derived from atomic clocks. For scientific purposes, quartz clocks were superseded by astonishingly accurate atomic clocks
Quartz time switch	A quartz time switch has a quartz movement similar to those used in quartz clocks. Switches are solid state so there are no moving parts. Quartz time switches are still used for domestic purposes
Reduction gear	A reduction gear is the counter train of an electric clock
Reluctance synchronous motor	A reluctance synchronous motor is synchronous motor in which the rotor is a temporary magnet
Rewind train	A rewind train is the train of gears used to rewind a mainspring automatically
Rotor	The rotor is the rotating part of a synchronous motor
RPM	RPM is the usual acronym for minimum retail price maintenance
rpm	Revolutions per minute

Stator	The stator is the stationary part of a synchronous motor
Striking train	A striking train is the train of gears that drives the striking work in a clock
Subsidiary dial	An indicating device with a dial and hands that is driven by electrical impulses from a master clock
Synchronous clock	A synchronous clock is a subsidiary dial driven by a synchronous motor
Synchronous motor	A synchronous motor is an electric motor driven by AC mains. Its speed is exactly related to the AC mains frequency, which in turn is controlled by a master clock at an electricity supply company
Synchronous time switch	A synchronous time switch is a time switch driven by a synchronous motor. It usually has a 24-hour dial. The switches are mechanical. Synchronous time switches were developed in parallel with synchronous electric clocks. Synchronous time switches are still used for domestic purposes, especially where heavy electrical loads need to be switched
Temporary magnet	A temporary magnet is made from a magnetically soft material which is temporarily magnetised by placing it within a magnetic field
Timekeeping element	A timekeeping element is an oscillator which operates at constant frequency. In a mechanical clock this is usually a pendulum, oscillating under gravity, or a balance wheel oscillating under a coiled hairspring. In a quartz clock it is a quartz crystal oscillating under the piezoelectric effect
Time switch	A time switch is used to operate a switch, or switches, at times specified by the user, and hence control electrical devices. A time switch usually has a visible time display as a complication
Wall clock	A wall clock is a clock that is intended for display high up on a wall. It has a dial that can be read from across the room

Chapter 1
Introduction

Alas! Regardless of their doom,
The little victims play;

Gray On a Distant Prospect of Eton College

Abstract This book is mostly about British domestic synchronous clocks made in the period 1930–1980, and their manufacturers. Here, British means a clock made in the UK for use on a UK mains supply (200–250 V, 50 Hz). Domestic means a portable clock with an analogue display suitable for use in a domestic setting. This introductory chapter has a wide scope, including the historical background to synchronous clocks, the AC mains supply, and user experience. It sets the scene for the main text in Part I (Chaps. 2, 3, 4, 5, and 6). British synchronous clock manufacturers and their brands are described in Chap. 2. How a synchronous clock works is explained in Chap. 3. Chapter 4 is an analysis of synchronous clock case types and styles. Chapter 5 gives general advice on the servicing of synchronous clocks. Part I ends with Chap. 6. This is an analysis of the marketing and reliability of synchronous clocks. It leads to an explanation of the rise and fall of synchronous clock technology. Part II is an organised database on British domestic synchronous clocks. Chapters 7, 8, 9, and 10 are galleries of mantel, bedside, wall, and granddaughter synchronous clocks respectively. There are back and front views of each clock, together with a brief description. Chapter 11 is a gallery of synchronous movements. It includes views of movements partly dismantled. There is a technical description of each movement.

1.1 Outline

In this book *British* means a clock made in the UK for use on a UK mains supply (200–250 V, 50 Hz). *Domestic* means a portable clock with an analogue display suitable for use in a domestic setting. For domestic use a clock should have an attractive appearance, an easily read dial, and keep good time to, say, within 2 min per week. There are several stakeholders in domestic synchronous clocks. Users are interested in their timekeeping, appearance, cost and reliability. Manufacturers are interested in brands, and in their production costs, marketing, detailed technical

L.P. Pook, *British Domestic Synchronous Clocks 1930–1980*, History of Mechanism and Machine Science 29, DOI 10.1007/978-3-319-14388-0_1

aspects, and patents. Regulatory authorities are interested in trading standards, safety, and the control of the AC mains supply.

This book is mostly about British domestic synchronous clocks made in the period 1930–1980, and their manufacturers. This introductory chapter has a wide scope, including the historical background to synchronous clocks, the AC mains supply, and user experience. It sets the scene for the main text in Part I (Chaps. 2, 3, 4, 5, and 6). British synchronous clock manufacturers and their brands are described in Chap. 2. How a synchronous clock works is explained in Chap. 3, including some magnetic theory and its application to synchronous motors. Chapter 4 is an analysis of synchronous clock case types and styles. Chapter 5 gives general advice on the servicing of synchronous clocks, and is based on many years experience. Synchronous clock servicing usually has to be done on a do it yourself basis. Part I ends with Chap. 6. This is an analysis of the marketing and reliability of synchronous clocks that leads to an explanation of the rise and fall of the technology.

Part II is an organised database on British domestic synchronous clocks, selected from a much wider database on synchronous clocks that has been assembled over many years. Some of the data are from published sources, but many data were acquired by the examination of numerous synchronous clocks. Chapters 7, 8, 9, and 10 are galleries of mantel, bedside, wall, and granddaughter synchronous clocks respectively. There are back and front views of each clock, together with a brief description. Chapter 11 is a gallery of synchronous movements. It includes views of movements partly dismantled. There is a technical description of each movement.

The correct terminology associated with synchronous clocks has been controversial since soon after their invention (Anonymous 1932). The glossary included in this book is based on popular usage and horological trade jargon, updated to make it reasonably consistent.

1.2   What Is a Synchronous Clock?

The simple answer to the question 'What is a synchronous clock?' is that it is a clock powered by a synchronous motor whose speed of rotation is precisely determined by the frequency of the AC mains supply. The hands of a synchronous clock are driven by suitable gearing, called reduction gear. Synchronous clocks sometimes include striking or chiming work as what are known as complications.

A synchronous motor can equally well be used to operate electrical switches at times selected by the user. The resulting device is called a synchronous time switch. There is no clear distinction between domestic synchronous clocks and synchronous time switches for domestic use. It is sometimes convenient to regard a time switch as a complication in a synchronous clock. For example, a synchronous alarm clock is a synchronous clock with a time switch and an audible alarm as complications. A novelty clock is a clock that is primarily an ornament. In general, synchronous novelty clocks are not suitable for use as domestic clocks.

1.3 Historical Background

1.3.1 Mechanical Clocks

Traditional mechanical clocks are based on a mechanical oscillator, usually either a pendulum or a balance wheel. Mechanical pendulum clocks have been in use for centuries for scientific and domestic purposes, and have five essential features (Pook 2011). Firstly, a timekeeping element, its pendulum. Secondly, a power source, usually either a falling weight or a mainspring. Thirdly, a controlling device, its escapement. The escapement is usually connected to the pendulum by a crank called a crutch. The crutch has a fork at its lower end that engages the pendulum. In the escapement one tooth of the escape wheel (usually called the 'scape wheel) is released at regular intervals. Fourthly, a counter train, this is its going train of gears which drives the fifth requirement which is an indicating device. This is usually a dial with hands that indicate hours, minutes and sometimes seconds. A disadvantage of a pendulum clock is that is has to be set in beat so that it ticks evenly. A pendulum clock cannot be moved without disturbing its timekeeping.

To illustrate how a pendulum clock works details of the escapement of an Eighteenth Century longcase pendulum clock are shown in Fig. 1.1. This is called an anchor escapement, a reference to its shape (Pook 2011). Figure 1.2 shows a sketch of an anchor escapement. The escapements of most modern mechanical pendulum clocks are essentially the same, although there are detail differences. As the pendulum swings to and fro the pallets move to and fro allowing one tooth of the 'scape wheel to escape at a time. As a tooth moves it applies a mechanical impulse to the pendulum to keep it swinging. The going train moves intermittently so the wheels are crossed out by removing material to leave a rim connected by spokes to a central hub. This reduces weight and hence decreases inertia. The clock is driven

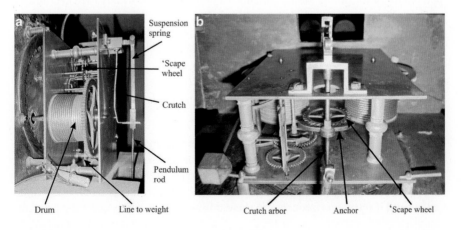

Fig. 1.1 Eighteenth Century longcase clock by John Barrow London, (**a**) movement, (**b**) anchor escapement

Fig. 1.2 Sketch of anchor escapement

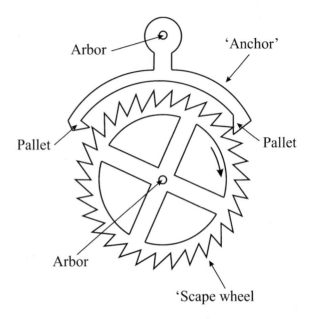

by a falling weight which drives a drum as the line to the weight unwinds from the drum. Forces at this end of the going train are high, but decrease as the escapement is approached, so the wheels are progressively lighter. Mechanical pendulum clocks were developed to a high level of precision for scientific purposes (Roberts 2003, 2004). Precision was further improved by the development of electrically driven pendulum clocks (Miles 2011).

Where clocks cannot be kept stationary, or where it convenient for them to be moved, they are fitted with a balance wheel as the timekeeping element and a mainspring as the power source. Balance wheel clocks were widely used in domestic settings. The movement of a balance wheel clock intended for use in a submarine is shown in Fig. 1.3. The mechanical oscillator consists of the balance wheel and coil spring. A lever communicates between the balance wheel and the 'scape wheel.

1.3.2 Synchronous Clocks

The idea of a synchronous clock, using a synchronous motor, was first suggested in 1887, and the first patent for a synchronous motor for use in clocks was applied for by Henry E Warren in 1916 (Philpott 1937). Warren produced his first synchronous clocks in the same year. The unique feature of a synchronous clock is that it is only useful if the mains frequency is standardised and accurately controlled, and the mains supply is reliable (Smith 1938). A general description of synchronous clocks was given by Ball (1931) who pointed out that there are three types. Firstly, clocks

Fig. 1.3 Movement of a
Vostok Watch Makers clock

|'Scape|Coil|Balance|
|wheel|spring|wheel|

which are self starting on being connected to the mains, and also after a power outage. Secondly, clocks which have to be started by hand on being connected, and also after an outage. Thirdly, clocks as in the first two types, but with an indicator to show that there has been an outage so the clock is showing an incorrect time. A synchronous clock can disturbed, for example during dusting, without affecting its timekeeping. However, if it is moved to a new location and plugged into a different socket it has to be reset.

Several books with extensive descriptions of particular makes of synchronous clocks have recently been published (Bird 2003; Lines 2012; Linz 2001, 2004; Miles 2011; Smith 2008; Tran Duy Li 1997). These books include descriptions of the principles of operation of synchronous clocks. However, they do not discuss the design of the cases, dials and hands of synchronous clocks, and their artistic context. Earlier, more general, books sometimes have sections devoted to synchronous clocks, with emphasis on movements (Britten 1978; Guye and Bossart 1957; Philpott 1935; Robinson 1942).

Some synchronous clocks have patent numbers on them, and patents are a useful source of information on synchronous movements. Details of patents and abstracts can be accessed at http://worldwide.espacenet.com/ (van Dulken 2013). To search for a British Patent add the prefix GB to the number: for example for Patent number 366710 enter GB366710.

1.3.3   Quartz Clocks

In a quartz clock the mechanical oscillator is a quartz crystal. This is driven by an electronic circuit via the piezoelectric effect. Quartz clocks for retail sale were first produced in 1971 (Smith 2008). Following advances in semiconductor engineering (Lojek 2007) battery powered quartz clocks for domestic use are now ubiquitous. They are cheap, reliable and accurate, and can be moved without affecting the time keeping. Some quartz clocks have electrically driven pendulums, these mimic mechanical clock pendulums for aesthetic reasons, but do not control timekeeping.

1.3.4   Atomic Clocks

An atomic clock is based on structural transitions which takes place within an individual atom (Audoin and Guinot 2001). In particle physics terms, transitions are between the atom's ground state and an excited state. These produce electromagnetic radiation. For a particular transition within a particular type of atom the frequency of the electromagnetic radiation is constant. Similar operations take place in all types of atomic clock. Conditions are such that the electromagnetic radiation is due to a particular transition. This is detected and electronic means are used to produce practically applicable signals.

1.3.5   Timelines

Broad timelines for domestic clocks, time switches for domestic use, and clocks for scientific purposes are instructive. Domestic clocks were originally mechanical. When domestic synchronous clocks were introduced they did not supersede mechanical clocks but were produced in parallel, often by manufacturers who also produced mechanical clocks (Bird 2003; Linz 2004; Robinson 1942; Smith 2008; Tran Duy Li 1997). However, when domestic quartz clocks were introduced they superseded synchronous clocks, but only partly superseded mechanical clocks. Time switches were originally mechanical, and not intended for domestic use. Synchronous time switches for domestic use were developed in parallel with synchronous clocks. When quartz time switches were introduced, they did not supersede synchronous time switches. Both types are still produced. Clocks for scientific purposes were originally mechanical. These were superseded by quartz clocks, and then by atomic clocks, always for improved precision (Audoin and Guinot 2001). Synchronous clocks were never used for scientific purposes.

1.4   AC Mains Supply

Standardised and accurately controlled frequencies for the AC mains supply were not introduced in the UK to make a time service possible, but in order to make interlinkage of generator outputs in a grid practical, and so enable systems of electrical load sharing (Stott 1946). The availability of a time service for use in synchronous electric clocks was a fortuitous by-product. Electricity supply companies would not have been interested in making special arrangements for devices which use very little electricity. Standardised frequencies are usually either 50 Hz, including the European Union, or 60 Hz, including USA and Canada. According to an advertisement Everett Edgcumbe & Co Ltd, in 1929, introduced the idea of using the public electricity supply in the UK for timekeeping (Anonymous 1940).

Soon after the 1914–1918 war, the UK Parliament appointed Electrical Commissioners who established the Central Electricity Generating Board, with instructions to co-ordinate the UK's electricity undertakings (Hope-Jones 1940). They evolved the Grid, which came into existence in 1927, and by 1933 had covered the whole country with a series of spider's webs whose centres were linked together. This was a network of mains carrying alternating current at an accurately controlled frequency of 50 Hz. At the time the frequency was standardised the voltage in the UK varied, and many British synchronous clocks are marked with a range of voltages, usually 200–250 V. Some clocks are marked with voltages within this range, and occasionally with a single voltage. In the UK the voltage has been standardised at a nominal value of 240 V for over 50 years. Throughout the European Union it is now 230 V +10 % −6 %. This means that British synchronous clocks can still be used in the UK. As an example, Fig. 1.4 shows a Smith *Gibraltar* synchronous Westminster chime granddaughter clock in use in the home office in which this book was written.

In 1932 it was noted (Ball 1932) that frequency controlled mains 'will soon become universal in the UK'. However, by the end of 1934 only half the number of homes in the UK was wired for mains electricity and of these nearly 20 % were on DC (Read 2012). An additional requirement is conveniently placed electrical sockets (AC outlets). In 1936 it was noted by Moore (1936) that ' . . . the synchronous-motor clock has not yet come into really general use. . . . an obstacle (perhaps the only important one) is the lack of suitable – i.e. safe and continuously-fed wiring points in the average house'.

Some houses in the UK had dedicated sockets (AC outlets) intended for synchronous clocks (Fig. 1.5). These are British Standard 3 pin 2 A sockets, which are still available, together with matching plugs. The power consumption of a synchronous clock is low, and is typically 3 W (Ball 1931). Dedicated sockets are therefore on lighting circuits. Special purpose 2 pin 1 A fused plugs and sockets were available for use with wall clocks. An example made by MK is shown in Fig. 1.6. The surface mounted socket is 2-1/8 in. (54 mm) square and the plug is held in place by a thumb screw. Its low profile means that it can be concealed behind a clock.

Fig. 1.4 Smith *Gibraltar*
synchronous Westminster
chime granddaughter clock,
height 137 cm

Fig. 1.5 A 3 pin 2 A socket
(AC outlet, *arrowed*) for a
synchronous clock on a
chimney breast of a house
built in the 1950s, and an Art
Deco Temco mantel clock
with pale translucent marble
and chrome case

Fig. 1.6 A 2 pin 1 A fused plug and socket made by MK

Statistics published in 1937 showed that interruptions to the mains electricity supply averaged one interruption per customer per year (Read 2012). Recent personal experience shows that this level has not been maintained. Typically, there are now two or three interruptions per year. In addition, there are local interruptions due to circuit breakers blowing.

A check during October 1953 showed timekeeping errors for a synchronous clock ranging from −49 to +15 s (Anonymous 1953). Errors are not cumulative so this was satisfactory for domestic purposes. Frequency control is still satisfactory and synchronous clocks in permanent use only need resetting following a power outage and when the hour changes for daylight saving.

1.5 User Experience

From the viewpoint of a UK owner synchronous clocks were a completely new technology whose heyday lasted about 50 years. The popularity of synchronous clocks was due to the fact that, providing the electricity supply was not interrupted, they kept good time, did not need regular attention and, like balance wheel mechanical clocks, could be disturbed without affecting the timekeeping. Some synchronous clocks are very reliable, and can run for decades without attention. This was predicted in 1938 when it was suggested that good quality synchronous clocks had an estimated life of 40 years (Anonymous 1938). Some models are self starting, and are therefore plug and run.

Synchronous clocks are becoming increasing popular with collectors. The best of them are works of art in their own right and, combined with good timekeeping, this makes them good clocks for domestic use. One of the attractions is that, compared with collectable mechanical clocks, they are still relatively cheap. High accuracy is not needed for domestic clocks or for time switches used for domestic purposes.

This is because very accurate time signals have been readily available for many years via radio transmissions and more recently on the Internet. This makes it easy to set clocks to time.

English law on the use of electrical appliances, including synchronous electric clocks, in private homes is less restrictive than is commonly believed (Wikipedia 2014). The Occupier's Liability Act 1957 imposes upon the occupier a duty of care. The occupier must '...take such care as in all the circumstances of the case is reasonable to see that the visitor will be reasonably safe in using the premises for the purposes for which he is invited or permitted by the occupier to be there.' The standard of care an occupier is expected to meet is the standard of a reasonable occupier. This is no different from the usual common law negligence standard of care. The Occupiers' Liability Act 1984 makes the scope of duty to a trespasser much narrower. In practice this means that collectors do not need to ensure that their synchronous clocks meet current electrical regulations.

References

Anonymous (1932) Synchronous motor clocks and electrical impulse dials. Horol J 75(889): 208–209

Anonymous (1938) Discussion at the British Horological Institute on the relative merits of synchronous electric and mechanical clocks January 12th, 1938. Horol J 80(953):6, 8, 10, 12, 14, 16–18

Anonymous (1940) Everett Edgcumbe. Horol J (82)984:10

Anonymous (1953) Synchronous timekeeping. Horol J 95(1143):826, Corrigendum. Tremayne, A (1953) Postcard

Audoin C, Guinot B (2001) The measurement of time. Time, frequency and the atomic clock. Cambridge University Press, Cambridge

Ball AE (1931) Electric clocks which operate on service mains. Chapter I. Horol J 73(870): 102–106

Ball AE (1932) Electric clocks which operate on service mains. Chapter X. Horol J 74(883): 109–111: 135

Bird C (2003) Metamec. The clockmaker. Antiquarian Horological Society, Dereham

Britten FJ (1978) The watch & clock makers' handbook, dictionary and guide, 16th edn. Revised by Good, R. Arco Publishing Company Inc, New York

Guye RP, Bossart M (1957) Horlogerie électrique, 2nd edn. Scriptar SA, Lausanne

Hope-Jones F (1940) Time from the grid. Horol J 82(984):5–11

Li TD (1997) New Haven clocks and watches. Arlington Book Co, Fairfax

Lines MA (2012) Ferranti synchronous electric clocks. Paperback edition with corrections. Zazzo Media, Milton Keynes

Linz J (2001) Electrifying time: telechron and GE clocks. Schiffer Publishing Co., Atglen

Linz J (2004) Westclox electric. Schiffer Publishing Co., Atglen

Lojek B (2007) History of semiconductor engineering. Springer, Berlin

Miles RHA (2011) Synchronome. Masters of electrical time keeping. The Antiquarian Horological Society, Ticehurst

Moore GE (1936) Miniature synchronous motors in service. Horol J 78(933):3–4

Philpott SF (1935) Modern electric clocks, 2nd edn. Sir Isaac Pitman & Sons Ltd., London

Philpott SF (1937) Modern synchronous clocks. 1. History of the synchronous clock. Horol J 79(46):3

Pook LP (2011) Understanding pendulums. A brief introduction. Springer, Dordrecht

Read D (2012) The synchronous mains revolution and 'Time gentlemen, please'. Antiquarian Horol 33(4):487–492

Roberts D (2003) Precision pendulum clocks: 300 year quest for accurate timekeeping in England. Schiffer Publishing, Atglen

Roberts D (2004) Precision pendulum clocks: France, Germany, America, and recent advancements. Schiffer Publishing, Atglen

Robinson TR (1942) Modern clocks. Their repair and maintenance, 2nd edn. N A G Press Ltd., London

Smith SJ (1938) 'Jim and the jingoes' XI. Horol J 81(961):23–28

Smith B (2008) Smiths domestic clocks, 2nd edn. Pierhead Publications Limited, Herne Bay

Stott HT (1946) Electricity and horology. Horol J 88(1048):23–30

van Dulken S (2013) What patents are, and how to find them. Electrical Horology group paper no. 85. Antiquarian Horological Society, Ticehurst

Wikipedia (2014) http://en.wikipedia.org. Accessed 2014

Part I
Main Text

Chapter 2
Manufacturers and Brands

Abstract Members of the Synchronous Clock Conference, which was founded in 1932, dominated the manufacture of domestic synchronous clocks in the UK up to the start of the Second World War in 1939. Production continued during the war, but on a reduced scale. After the end of the war in 1945 big names re-reappeared, including Ferranti Ltd, Smiths Industries, and Telephone Manufacturing Co Ltd. There were many new UK manufacturers of domestic synchronous clocks, but only one of these, Metamec, survived to become a major manufacturer. Two American companies set up factories to produce synchronous clocks in the UK. Westclox survived to become a major manufacturer. Information, but not detailed company histories, on manufacturers of synchronous clocks and synchronous movements featured in Chaps. 7, 8, 9, 10, and 11 is given in this chapter. Use of brands on clocks, including company names, is often inconsistent and confusing. A list of brands used to identify synchronous clocks in Chaps. 7, 8, 9, and 10 is given, together with the owner of each brand, where known. In the UK manufacture of synchronous clocks became widespread in the 1930s with the availability of mains electricity of standardised and accurately controlled frequency. When an old synchronous clock is acquired it is not usually accompanied by any manufacturer's paper work, and detailed identification may be difficult. Identification of synchronous clocks as definitely British made is usually straightforward.

2.1 Introduction

Members of the Synchronous Clock Conference (Anonymous 1932), which was founded in 1932, dominated the manufacture of domestic synchronous clocks in the UK up to the start of the Second World War in 1939. There are one make books on three of these companies: Ferranti Ltd (Lines 2012), Smiths Industries (Smith 2008) and Synchronome Company (Miles 2011). There were numerous other manufactures but relatively little information is available about these. Production continued during the war, but on a reduced scale.

The War ended in 1945. There is a section on synchronous clocks in the report of a lecture given in March 1946 (Barrett 1946). It includes the following:

> That these clocks have a very bright future there is no doubt. In fact they are the clocks of
> the present and the future as far as can be visualise. In future, British manufacturers will

© Springer International Publishing Switzerland 2015

L.P. Pook, *British Domestic Synchronous Clocks 1930–1980*, History of Mechanism and Machine Science 29, DOI 10.1007/978-3-319-14388-0_2

make 12-hour ordinary and 24-hour automatic synchronous electric alarms: timepieces for every conceivable purpose and in every possible style; strike and chimes – and what a boon it will be not having to wind heavy springs or having to worry about setting the clock in beat, or to trouble about stopping the clock for dusting; wall clocks for every purpose – industrial and domestic; clocks for switching on the radio at a pre-determined time, switch on and off the electric cooker. In fact these clocks will revolutionise the horological field. In regard to production, all the well known makes will re-appear and many new makes – seven more to my knowledge.

Some of the predictions were correct. Big names did re-reappear and time switches did appear. There were many new UK manufacturers of synchronous clocks, but only one of these, Metamec (Bird 2003), survived to become a major manufacturer. Two American companies set up factories to produce synchronous clocks in the UK. These were the New Haven Clock Company (Li 1997) and Westclox (Linz 2004). The prediction that synchronous clocks would revolutionise the horological field turned out to be over optimistic.

Information, but not detailed company histories, on some of the manufacturers of synchronous featured in Chaps. 7, 8, 9, and 10 is given in this chapter. Use of brands on clocks, including company names, is often inconsistent and confusing. A list of brands used to identify clocks and movements in Chaps. 7, 8, 9, 10, and 11 is given in Sect. 2.2, together with the owner of each brand, where known.

2.1.1   Trade Associations

In early 1932 the British Watch and Clock Makers Association was formed in order to agree minimum retail prices, and to lobby for import duties on foreign imports (Lines 2012; Nye 2014). The Synchronous Clock Conference, later known as the British Synchronous Clock Conference, was formed in November 1932 to foster the interests of UK synchronous clock manufacturers (Anonymous 1932). The founding members were: British Sangamo Co Ltd, English Clock & Watch Manufacturers Ltd, Synclocks Ltd, Ferranti Ltd, Smith's English Clocks Ltd, Synchronome Co Ltd, and Telephone Manufacturing Co Ltd. Lines (2012) gives a list of founding members which has some differences, but these are due to the use of different brands owned by the same company. The Conference's objectives included: to popularise the use of synchronous electric clocks, to standardise synchronous electric clocks as far as possible in many important details, and to see that existing and new houses are provided with suitable wiring. The Conference was also a cartel to control prices. In 1936 an agreement with Approved Wholesalers was reached to ensure that; retail prices were rigidly adhered to, and retail discounts maintained (Anonymous 1936). In 1937 the Conference increased prices by 10 % with effect from 8 Mar 1937 (Anonymous 1937c). This was the first increase since the formation of the Conference. Previously (December 1931) Smith, Ferranti, Sangamo and Everett Edgecumbe had agreed on a minimum retail price of 30/- (Lines 2012). This was a lot of money. In the 1930s: wage rates were about 1/- per hour (eHow 2014).

The 1936 agreement also provided that all clocks made by members of the Conference shall pass the tests imposed by the British Standards Institution. A review article emphasised the importance of electrical safety (Smith 1937), and in 1937, the British Horological Institute, in association with the British Clock Manufacturers' Association and the British Synchronous Clock Conference, was making arrangements for official testing of synchronous clocks to British Standard 472, issued in 1932 (Anonymous 1937d). A revised version of BS 472 was issued in 1962 (Anonymous 1962). The current standard is BS EN 60335-2-26:2003+A1:2008 (Anonymous 2003).

2.1.2 Synchronous Clock Production

In the UK, the manufacture of synchronous clocks, including synchronous alarm clocks became widespread in the 1930s with the availability of mains electricity of standardised and accurately controlled frequency. UK output of synchronous clocks was 5,000 in 1930, 19,000 in 1931 and 183,000 in 1932 (Nye 2014). Corresponding figures for mechanical clocks were 24,000, 36,000 and 636,000. Annual supplies of electric clocks to the UK market, including imported clocks were approximately (Anonymous 1951): 1935: 108,000, 1945: 192,000, 1946: 732,000, 1947: 924,000, 1948: 576,000, 1949: 432,000. Most of these would have been synchronous electric clocks. In 1935 electric clock production was around 100,000 and in 1937 around 165, 000 (Lines 2012). In 1938 about 500,000 synchronous clocks were produced in the UK (Barrett 1946). During the Second World War (1939–1945) production was reduced (Anonymous 1940a, 1945b). In 1935 the total number of clocks produced was about 900,000 (Smith 1937) so in that year the number of synchronous clocks produced was about one tenth of the total. Production continued on a large scale until the 1960s. In July 1945 Ferranti announced a synchronous clock production programme for August 1945: 3,000, September 1945: 4,000, October 1945: 5,000, November 1945: 6,000, and ultimately 10,000 per month (Lines 2012). 4,281 clocks were produced in September 1945. In 1947 Ferranti planned to make 150,000 synchronous clocks and 50,000 synchronous alarm clocks. In 1937 and 1939 it was noted that clocks are made with a wide range of cases, in a variety of materials, in the belief that this stimulated sales (Anonymous 1939a; Pook 2014; Seager 1937).

2.1.3 Identification of Synchronous Clocks

When an old synchronous clock is acquired it is not usually accompanied by any manufacturer's paper work, and detailed identification may be difficult.

Identification of synchronous clocks as British made is straightforward since they are marked as such somewhere on the clock. The usual marking are 'Made in England', 'Made in Scotland' or 'British made'. This is usually reliable since it was illegal for a retailer to sell as British made a foreign made clock, or a clock containing foreign made parts. This was enforced by the courts. However, there are some later clocks marked as British made which are fitted with foreign made movements. These can usually be detected by examining the movement. Clocks not marked as British made are nearly always foreign made, or contain foreign parts. Manufacturers' markings are sometimes misleading in that they give the impression that a clock is British made without actually saying so.

Markings on a clock are a useful source of information, but interpretation may be needed. These include brands which appear on the front and back of a clock, model names or numbers which appear on the back, and patent numbers on the back of the clock, or on the movement. The clock usually has to be dismantled to see patent numbers on the movement. If paperwork or a box is available they may enable the manufacturer or model to be identified. Published information on synchronous clocks, especially one make books, often include the information needed to identify a clock. In the owner of a brand is known then it is usually possible to identify the manufacturer. If all else fails, then stylistic considerations will sometimes provide clues. All these sources of information have been used to identify clocks described in Chaps. 7, 8, 9, and 10. When more than brand appears on a clock, and they do not all belong to same manufacturer, it is usually possible to work out why. When there are no markings on a clock case to show by whom it was made, it is usually safe to assume that it was made, or at least commissioned, by the manufacturer of the synchronous movement.

Manufacturing practices mean that determining the actual manufacturer of a clock and its movement may not be straightforward. There are several possible reasons. Rebranding may have attributed the clock to a different manufacturer, or to a retailer. Jewellers often sold own brand goods. The movement may have been supplied by a different manufacturer, and may or may not have been rebranded. An apparently rebranded or unbranded movement might have been made under license. Occasionally, patents belonging to another manufacturer are acknowledged and therefore provide a clue. Some clocks are marriages. These are usually clocks with insertion movements in which the original movement has been replaced with a later movement, presumably because the original movement failed. This was sometimes done by Metamec when a clock was sent to them for servicing (Bird 2003). Later replacements can often be detected by anachronisms or stylistic considerations. Sometimes, a movement from another clock is installed in a homemade case, or the case has been heavily modified. This can usually be detected from the method of construction or by stylistic considerations.

2.2   Brands

Brands owned by manufacturers that appear on domestic synchronous clocks and synchronous time switches for domestic use are of three types: brands that are trademarks, brands used for marketing, and brands that are company names.

The categories sometimes overlap. To avoid confusion some of the brands that appear on clocks have been chosen for use as section titles etc. in Chaps. 7, 8, 9, and 10, and some manufacturers names chosen for the company information section titles in this chapter. Brands that appear on clocks illustrated in Chaps. 7, 8, 9, and 10, and corresponding manufacturers, are shown Table 2.1. The brands that appear on clocks are in alphabetical order so the table can used as an index to company information in Sects. 2.3.1, 2.3.2, 2.3.3, 2.3.4, 2.3.5, 2.3.6, 2.3.7, 2.3.8, 2.3.9, 2.3.10, 2.3.11, 2.3.12, 2.3.13, 2.3.14, 2.3.15, 2.3.16, 2.3.17, 2.3.18, 2.3.19, 2.3.20, and 2.3.21.

2.3   Manufacturers

Information is restricted to the manufacturers of domestic synchronous clocks and synchronous time switches for domestic use described in Chaps. 7, 8, 9, and 10. It was impossible to find useful information on some manufacturers. The notes in Sects. 2.3.1, 2.3.2, 2.3.3, 2.3.4, 2.3.5, 2.3.6, 2.3.7, 2.3.8, 2.3.9, 2.3.10, 2.3.11, 2.3.12, 2.3.13, 2.3.14, 2.3.15, 2.3.16, 2.3.17, 2.3.18, 2.3.19, 2.3.20, and 2.3.21 are intended to set the scene for the manufacture of synchronous clocks and synchronous time switches, rather then to provide detailed company histories. Synchronous clock manufacture was sometimes a company's only activity, sometimes a major part of its activities, and sometimes a minor part of its activities. A noticeable feature is the widespread use of synchronous clock movements made by other manufacturers.

Companies that started making domestic synchronous clocks and synchronous time switches for domestic use in the 1930s included British Vacuum Cleaner & Engineering Co Ltd, Ferranti Ltd, Garrard Clocks Ltd, General Electric Co Ltd, Horstmann Gear Co Ltd, Ismay Industries, Sangamo Weston Ltd, Smiths Industries, Synchronome Co Ltd, and Telephone Manufacturing Co. Of these Ferranti Ltd, Sangamo Weston Ltd, Smiths Industries, Synchronome Co Ltd, and Telephone Manufacturing Co were members of the Synchronous Clock Conference (Sect. 2.1.1). Apparent discrepancies are due to the use of different brands.

Companies that started making domestic synchronous clocks in the 1940s, after the end of the Second World War in 1945, included Alexander Clark Co Ltd, Clyde Clocks, Elco Clocks and Watches Ltd, Ingersoll Ltd, Jentique Ltd, Metalair Ltd, and Westclox Ltd, with the probable addition of Camerer, Kuss & Co, Edison Swan Electric Co, and Franco-British Electrical Co Ltd. Smiths Industries started making synchronous time switches for domestic use in the 1930s. Horstmann Gear Co Ltd and Jentique Ltd started making them after the end of the Second World War in 1945.

Distribution, sometimes called marketing, was usually made the responsibility of a separate marketing company, which was sometimes a subsidiary of the manufacturer. For example, the Telephone Manufacturing Co used T.M.C.-Harwell (Sales) to distribute Temco Clocks (John Harris personal communication 2014).

Table 2.1 Brands that appear on clocks and corresponding manufacturers

Brand on clock	Manufacturer
Alexander Clark	Alexander Clark Co Ltd.
BEM	Sangamo Weston Ltd
Bowden & Sons	Unknown
British Electric Meters Ltd.	Sangamo Weston Ltd
Camerer Kuss	Camerer Kuss & Co.
Clyde	Clyde Clocks
Ediswan	Edison Swan Electric Co
Elco	Elco Clocks and Watches Ltd
Empire	Smiths Industries
English Clock Systems	Smiths Industries
Ferranti	Ferranti Ltd
Franco-British Electrical Co Ltd	Franco-British Electrical Co Ltd
Garrard	Garrard Clocks Ltd
GEC	General Electric Co Ltd
Genalex	General Electric Co Ltd
Goblin	British Vacuum Cleaner & Engineering Co Ltd
Horstmann	Horstmann Gear Co
Ingersoll	Ingersoll Ltd
Ismay	Ismay Industries
Kelnore	J H Jerrim & Co Ltd, Birmingham
Liberty	Liberty plc
Magneta	British Vacuum Cleaner & Engineering Co Ltd
Marigold	Marigold
Metalair	Metalair Ltd
Metamec	Jentique Ltd
Riley	Riley
Sangamo	Sangamo Weston Ltd
S Smith and Sons (England) Ltd	Smiths Industries
SEC	Smiths Industries
SECTRIC	Smiths Industries
Smith	Smiths Industries
Smith Electric	Smiths Industries
Smiths	Smiths Industries
Smiths Clocks & Watches Ltd	Smiths Industries
Smiths English Clocks	Smiths Industries
Smiths English Clocks Ltd	Smiths Industries
Smith's English Clocks Ltd	Smiths Industries
Smiths Industries Limited	Smiths Industries
Sterling	Ismay Industries
Synchronomains	Synchronome Co Ltd
Synchronome	Synchronome Co Ltd
Synchronous Electric Clocks Ltd	Smiths Industries
Temco	Telephone Manufacturing Co Ltd
Westclox	Westclox Ltd

T.M.C.-Harwell (Sales) was founded by John Harris' grandfather following the realisation that Telephone Manufacturing Co had no expertise in the distribution of consumer goods. T.M.C.-Harwell (Sales) had specially fitted vans to display available Temco clocks (Anonymous 1949).

2.3.1 Alexander Clark Co Ltd

Alexander Clark Co, 125 and 126 Fenchurch Street, London, were established in 1890. In 1914 they had 620 employers and were silversmiths, cutlers, precious stone mounters and dressing bag manufacturers (Grace's Guide 2014). It is not clear when they became Alexander Clark Co Ltd, or when they started producing clocks. This was probably after the Second World War ended in 1945. They used synchronous clock movements made by another manufacturer.

2.3.2 British Vacuum Cleaner & Engineering Co Ltd

The British Vacuum Cleaner & Engineering Co Ltd was founded by Hubert Cecil Booth in 1902 and introduced the first vacuum cleaners to the UK market (Wikipedia 2014). In 1947 they were listed as manufacturers of Goblin controlled time domestic radios, incorporating a superhet radio receiver with a synchronous electric clock, and Magneta industrial radio receivers and public address systems. In 1961 they were listed as manufacturers of industrial and domestic vacuum cleaners, spin dryers, polishing machines, industrial power plant, washing machines, automatic teamaker, and central cleaning installations.

Goblin and Magneta synchronous clocks were made at the British Vacuum Cleaner & Engineering Co Ltd, Goblin Works, Leatherhead, Surrey. The factory is described by Anonymous (1939b). Goblin synchronous clocks, but not Magneta synchronous clocks, were displayed at the 1947 British Industries Fair (Anonymous 1947b). The Goblin trade mark was used from 1926. The Magneta Time Co Ltd was a division of the British Vacuum and Engineering Co Ltd. (ClockDoc 2014). They made their own synchronous clock movements.

2.3.3 Camerer, Kuss & Co

Advertisements from 1882 to 1899 show that Camerer, Kuss & Co, were watch and clock manufactures at 56 New Oxford Street, London (Grace's Guide 2014). In a 1947 advertisement the company name is Camerer, Cuss & Co, specialists in horology, and in a 1957 advertisement it is Camerer Cuss, makers of good clocks and watches since 1788. They used synchronous clock movements made by another manufacturer.

2.3.4   Clyde Clocks

Clyde Clocks, originally located at 142 West Nile Street, Glasgow, developed from the war time manufacture of aircraft components by British-Foreign Agencies Ltd. Faced with the need to convert to peace time production, and to utilise the machine tool capacity built up over the war period, the firm began the manufacture of synchronous electric clocks. In July 1947 it was the only firm in Scotland manufacturing synchronous clocks from start to finish. The firm moved to larger premises at 28–32 Graham Square, Glasgow (Anonymous 1947a; Grace's Guide 2014). Clyde Clocks were exhibited at the British Industries Fairs in 1947 (Grace's Guide 2014) and 1949 (Anonymous 1949). They used their own synchronous clock movements.

2.3.5   Edison Swan Electric Co

The history of the Edison Swan Electric Co is complicated and confusing. It was formed as the Edison and Swan Electric Light Company Limited on 26 October 1883 with the merger of the Swan United Electric Company and the Edison Electric Light Company (Grace's Guide 2014; Wikipedia 2014). Some of the information in these references is conflicting. In 1892 the company was merged into the General Electric Co, but retained its separate identity. After 1905 the company seems to have been known mostly as Edison Swan Electric Co. The Edison Swan Electric Co set up the first UK radio valve factory in 1916. In the 1920s it was acquired by the General Electric Co and continued to retain its separate identity (Sect. 2.3.10). Initially the Ediswan Electric Co was a manufacturer of incandescent light bulbs branded Ediswan. A vacuum cleaner was introduced in 1924. They later made radio valves, crystal radio sets, vacuum cleaners, car batteries, cathode ray tubes, electric cables, and flight information recorders.

Production of synchronous clocks under the Ediswan brand appears to have started after the end of the Second World Was in 1945. They used synchronous clock movements made by another manufacturer.

2.3.6   Elco Clocks and Watches Ltd

Elco synchronous clocks were made by Elco Clocks & Watches Ltd, Elco House, 51 Hatton Garden, London. They were researching a synchronous movement in

1945 (Anonymous 1945a). Also in 1945 they took an extensive factory in Wales, planning to produced a wide range of synchronous clocks (Anonymous 1945b). Elco synchronous clocks were exhibited at the 1947 British Industries Fair (Anonymous 1947b). They were advertised in 1947 (Anonymous 1947c). They used their own synchronous clock movements.

2.3.7 Ferranti Ltd

The history of Ferranti Ltd, Hollinwood, Manchester, who made a wide range of electrical products is complicated (Lines 2012). The Museum of Science and Industry (MOSI), Manchester, England holds the Ferranti collection which consists of the Ferranti archive and about 1400 objects from 1882 to 1993 when the company stopped trading. The company was founded in 1882 by Sebastian Zani de Ferranti as Ferranti, Thompson and Ince Ltd.

In 1930 Sir Vincent de Ferranti became chairman of Ferranti Ltd and he initiated synchronous clock production within the Domestic Appliances Department. They produced synchronous clocks from 1932 to 1957. In 1932 they aimed to make just under 3,000 synchronous clocks per week, together with 1,000 synchronous alarm clocks per week. Ferranti Ltd used their own synchronous clock movements, and also used synchronous clock movements made by another manufacturer. They also supplied synchronous clock movements to other manufacturers.

2.3.8 Franco-British Electrical Co Ltd

In 1914 Franco-British Electrical Co was listed as an electrical goods manufacturer, supplier or engineer, and as a manufacturer of electric signs (Grace's Guide 2014). In 1929 their product range included electric letter signs, illuminated facias and enamelled plates. In 1962 their address was 25 Oxford Street, London, W1. Production of synchronous clocks brand appears to have started after the end of the Second World Was in 1945. They used synchronous clock movements made by another manufacturer.

2.3.9 Garrard Clocks Ltd

Garrard and Company was established in 1721 by George Wickes and they were appointed Crown Jewellers of London (Grace's Guide 2014). They had an international reputation for craftsmanship in the design and manufacture of jewellery, gold and silverware. They were incorporated as a limited company in 1909. Specialities were fine diamond and coloured gem jewellery and pearls, gold and silver plate, artistic productions in both gold and silver, and the Insignia of all the principal Orders of Knighthood. In 1915 Garrard and Company Ltd formed the Garrard Engineering Manufacturing Company Ltd with Major S H Garrard as Chairman, to run a factory set up to create precision rangefinders, as they had the specialist equipment necessary. The company became a separate entity following the death of Major Garrard in 1945.

A subsidiary of the Garrard Engineering Manufacturing Company Ltd, Garrard Clocks Ltd, 117 Golden Lane, London was set up in 1931 to manufacture synchronous and mechanical clocks. Production of clocks continued until 1954 (Grace's Guide 2014; Anonymous 1940b). Production was discontinued during the Second World War and resumed in 1945 (Anonymous 1945a). They used their own synchronous clock movements.

2.3.10 General Electric Co Ltd

The General Electric Company (GEC) was a major British-based industrial conglomerate, involved in consumer and defence electronics, communications and engineering (Wikepedia 2014). It is not to be confused with the American company General Electric (GE), who also made synchronous clocks (Linz 2001). GEC had its origins in G. Binswanger and Company, an electrical goods wholesaler established in London in the 1880s by Gustav Binswanger (later Gustav Byng). In 1886 Hugo Hirst joined Byng, and the company changed its name to The General Electric Apparatus Company (G. Binswanger). This small business prospered, and the following year the company acquired its first factory in Salford, England where telephones, electric bells, ceiling roses and switches were manufactured. In 1889, the business was incorporated as a private company known as General Electric Company Ltd The company was expanding rapidly, and trading in 'Everything Electrical', a phrase that was to become synonymous with GEC. In 1900 GEC was incorporated as a public limited company The General Electric Company (1900) Ltd, (the 1900 was dropped 3 years later). The outbreak of the First World War in 1914 transformed GEC into a major player in the electrical industry. Between 1914 and 1939, GEC expanded to become an international corporation and a national institution.

In the 1920s the company was heavily involved in the creation of the UK National Grid so the manufacture of synchronous clocks in the 1930s was a natural development. However, it is not clear whether and to what extent GEC manufactured or assembled synchronous clocks, or simply marketed clocks, initially under the GEC brand, and later the Genalex brand. Major re-organisation and rationalisation in the early 1960s appear to have coincided with the end of synchronous clock production. GEC and Genalex synchronous clocks are fitted with movements made by another manufacturer.

2.3.11 Horstmann Gear Co Ltd

In the 1850s Gustav Horstmann developed a thriving retail and clock making business which became G. Horstmann and Sons (Grace's Guide 2014). In 1902 a clockwork gas controller was introduced for the automatic control of street lighting. In 1904 Sidney Horstmann and his brothers established the Horstmann Gear Co in 1904 to develop a variable speed gear-box he had invented for cars and motorcycles. The gearbox was not a success, but the company became well-known for its clockwork mechanisms and timers. The brothers patented the Solar Dial which automatically adjusted lighting times at dusk and dawn throughout the year. It was the start of nearly 80 years of Horstmann's manufacturing involvement in the street lighting controls market.

After the end of the First World War in 1918 the company developed new products including domestic clocks. Their main products were street lighting controls, gas ignition devices, time switches and gauges. The first system for timing central heating was introduced in 1939. Products during the Second World War included time switches for mines. Products introduced after the end of the war included time switches for domestic electric and gas-fired central heating and. They used their own synchronous time switch movements.

2.3.12 Ingersoll Ltd

The Ingersoll Watch Company grew out of a mail order business (R H Ingersoll & Bro) started in New York City in 1882 by 21-year-old Robert Hawley Ingersoll and his brother Charles Henry Ingersoll (Wikipedia 2014). The company initially sold low cost items such as rubber stamps. The first watches were introduced into the catalogue in 1892, supplied by the Waterbury Clock Company. In 1904 Ingersoll opened a store in London. In 1905 they introduced a pocket watch made by a British subsidiary, Ingersoll Ltd. Ingersoll Watch Company went bankrupt in 1921, but Ingersoll Ltd continued to trade at St John Street, London.

Ingersoll synchronous clocks were advertised for sale in 1934 (Anonymous 1934). After the end of the Second World War, in a joint venture with Smiths Industries Ltd and Vickers Armstrong, Ingersoll Ltd set up The Anglo-Celtic Company at factory on the Ynyscedwyn estate. This was on the outskirts of the village of Ystradgynlais, near Swansea, Wales. Vickers Armstrong later withdrew. The factory was officially opened in 1948 (Smith 2008). Ingersoll synchronous clocks do not appear to have been made at Ystradgynlais since all known examples are marked Made in England. They used synchronous clock movements made by another manufacturer.

2.3.13   Ismay Industries

Ismay Industries was formed in 1935 to take advantage of various opportunities in the electrical industry. They acquired the existing company John Ismay and Sons and expanded by further acquisitions (Grace's Guide 2014). Synchronous electric clock production was started in 1936 by a subsidiary, Sterling Clock Co Ltd, Dagenham, Essex, who used the brand Ismay (Anonymous 1937a). The Sterling Clock Co Ltd must not be confused with the Sterling Clock Company, Lasalle, Ill, USA. Ismay clocks had foreign made movements. Early in 1937 the Sterling Clock Co Ltd and Ismay Distributors were fined for selling Ismay synchronous clocks as British made when they had foreign made movements (Anonymous 1937e). The Ismay brand was still being used in August 1937 (Anonymous 1937b). In September 1937 the brand used had been changed to Sterling (Anonymous 1937b). Clocks branded Sterling had the same type of movement as Ismay clocks but were marked Made in England, so production of the movements must have been moved to Dagenham. The Sterling Clock Co. Ltd was still in existence in August 1938 (Anonymous 1938) but by 1947 it had become Sterling-Croydon Clocks, Pioneer House, Queensway, Waddon, Croydon, Surrey (Anonymous 1947b). They used their own synchronous clock movements.

2.3.14   Jentique Ltd

Jentique Ltd, Dereham, Norfolk had its origin in 1923 when Geoffrey Bowman Jenkins formed The Woodcraft Patents Company who made model boats and yachts (Bird 2003; Wikipedia 2014). In 1934 Jenkins Productions was formed to make dining and occasional furniture. Metamec Ltd was incorporated on 6 April 1941, and in 1942 the company name was changed to Jentique Ltd. Plans to make electric clocks were first considered in 1944. The Metamec trade mark was registered on 13 January 1947, and the first Metamec synchronous electric clocks were produced in the same year. Metamec was the clock making division of Jentique Ltd with its own identity. Metamec also made mechanical clocks and, later, quartz clocks. Between 1960 and 1970 production rose to 25,000 clocks per week and Metamec became the largest clock manufacturer in the UK. They were the most successful of the firms that started making synchronous clocks after the end of the Second World War in 1945. Jentique Ltd declined in the 1980s, and clock production stopped in 1984. In January 1985 the Metamec division was bought by F.K.I., and the rest of Jentique Ltd by Peter Black Holdings. Clock production was later restarted, and as of 2011 the Metamec name was still in use. Metamec used their own synchronous clock movements, and also supplied synchronous clock movements to other manufacturers.

2.3.15 Liberty plc

The company now known as Liberty plc was founded in 1975, as Liberty and Co, by Arthur Lasenby Liberty (Grace's Guide 2014; Wikipedia 2014). They occupied a shop in Regent Street, London. They moved into the newly built department store in Regent Street in 1924, and still occupy the store. In 1955 they began opening department stores in other UK cities. These were all closed in the 1990s, and small shops opened at airports. Since 1988, Liberty has had a subsidiary in Japan which sells Liberty-branded products in major Japanese shops.

The business was founded to sell ornaments, fabrics (for which they were to become especially famous) and miscellaneous objects d'art from the Far East, including Japan. From the 1890s work was commissioned from leading English designers and was sometimes sold under the Liberty brand. Many of these designers practised the artistic styles known as Arts and Crafts, and Art Nouveau. The store became one of the most famous in London. During the 1950s the tradition for fashionable and eclectic design was continued, with new designers involved. All departments in the store sold both traditional and contemporary designs. During the 1960s extravagant and Eastern influences became fashionable, as well as Art Deco, and Liberty adapted their designs. It is not clear when synchronous clock production started or for how long it continued.

2.3.16 Metalair Ltd

During the Second World War Metalair Ltd, Wokingham, Berks, made metal aircraft components. Advertisements in 1941 and 1943 offered their services (John Nuttall personal communication 2010). He believes that after the war Metalair diversified into other products, including synchronous clocks. Similarly, Malcolm Whatley (personal communication 2010), the son of the founder of Metalair, states that, after the war, contracts were few and far between, and to keep the business going Metalair manufactured electric clocks, anodised tea trays, children's tricycles, and the first rotary lawn mower called the Ladybird. Synchronous clocks with guarantee cards dated April and May 1947 are known. After the founder's death in 1956, the company had to be sold to pay death duties, and the company is believed to have moved to the Norfolk area. They used their own synchronous clock movements.

2.3.17 Sangamo Weston Ltd

British Sangamo Co Ltd was incorporated as a wholly-owned subsidiary of the Sangamo Electric Company, an American company in 1921 (Grace's Guide 2014). It became a public company in 1935. By 1937 British Sangamo Co Ltd had become

Sangamo Weston Ltd, Cambridge Arterial Road, Enfield, Middlesex. Its activities were expanded by acquisition of companies manufacturing electrical instruments and control systems. British Electric Meters Ltd was a Sangamo Western subsidiary. They used the brands British Electric Meters Ltd and BEM. They used their own synchronous clock movements. Synchronous time switches were branded Sangamo Western Ltd.

2.3.18 Smiths Industries

The history and company structure of Smiths Industries is complicated and confusing (Smith 2008; Grace's Guide 2014; Nye 2014). The company that became Smiths Industries, and is now known as Smiths Group plc, has its origin as a watch and clock business, S Smith and Son, founded by Samuel Smith in 1851. The company was later known as Samuel Smith, and then as Samuel Smith and Son. In 1882 it was listed as watch maker and jeweller, in 1884 as jeweller, optician, watchmaker and jet ornament manufacturer, in 1895 as watchmaker and jeweller, and in 1898 as watchmakers to the Admiralty, high-class watches with certificates from the Royal Observatory and jewellers. In 1900 they produced the first British milometer and speedometer for cars. S. Smith and Sons (Motor Accessories) was formed to take over from S Smith and Son the manufacture of speedometers and other motor accessories. A department to make escapements was formed in 1928. These had previously been imported from Switzerland. A trading deal was agreed in 1930 with Joseph Lucas Ltd whereby the two would not compete in certain areas and they became the dominant supplier of instruments to British motorcar and motorcycle firms.

S. Smith and Sons (Motor Accessories) re-entered the domestic clock market in 1931 and formed a new company, Smiths English Clocks, as the clock and watch division. They were one of the first companies to produce synchronous clocks in the UK, and became the dominant manufacturer. In 1932 they bought English Clock and Watch Manufacturers and acquired the brand names Astral and Empire which they used extensively. They bought Enfield Clock Co in 1934 and produced mechanical clocks under the Enfield name until about 1950. The striking and chiming work used in Enfield clocks was adapted for use in synchronous clocks.

In 1944 a major regrouping of the whole Smith organisation was carried out. The name of the principal company was changed from S Smith and Sons (Motor Accessories) Ltd to S. Smith and Sons (England) Ltd with four divisions. One of them was Smiths English Clocks, which included the following companies: All British Escapements Ltd, Enfield Clock Co. (London) Ltd, English Clock Systems Ltd, British Precision Springs Ltd, J E V Winterbourne Ltd, Pullers Instruments Ltd, United Kingdom Clock Co. Ltd, and Clock Components Ltd. It is not clear which of these companies made domestic synchronous electric clocks. English Clock Systems Ltd were only concerned with industrial clocks and supplied railways, schools, factories, shops and offices. Types supplied included floral clocks, street clocks, advertising clocks, and general purpose industrial clocks. They used the brand English Clock Systems.

After the end of the Second World War, in a joint venture with Ingersoll Ltd and Vickers Armstrong, Smiths Industries Ltd set up The Anglo-Celtic Company at factory on the Ynyscedwyn estate. This was on the outskirts of the village of Ystradgynlais, near Swansea, Wales. Vickers Armstrong later withdrew. The factory was officially opened in 1948 (Smith 2008). In 1965 with increasing diversification and international operations the name Smiths Industries was adopted to reflect wider operations and in 1966 the name of the principal company of the Smiths group was changed from S. Smith and Sons (England) Ltd to Smiths Industries. In 1979 they stopped making clocks and watches. They used their own synchronous clock movements, and sometimes used synchronous clock movements made by another manufacturer. They also supplied synchronous clock movements to other manufacturers.

2.3.19 Synchronome Co Ltd

Synchronome Co Ltd had its origin in 1895 when Synchronome of London was established by Frank Hope-Jones (Miles 2011; Grace's Guide 2014). After a turbulent history it was incorporated as the Synchronome Co Ltd in February 1912 with Frank Hope-Jones as one of the directors. Further turbulence followed the end of the Second World Was in 1945 and in 1969 Synchronome was merged with the long established safe maker John Tann to form Tann-Synchronome. Ltd. This was not a commercial success and on 2 May 1988 it was acquired by the Tunstall Group. The business, assets and liabilities were transferred to Tunstall Telecom ltd on 30 September 1988. Most of the business was closed down.

The Synchronome Co was best known as a manufacturer of electric master clocks and subsidiary dials. The company started producing synchronous clocks in 1931 Early synchronous clocks were branded Synchronomains and later synchronous clocks were branded Synchronome. They used synchronous clock movements made by other manufacturers.

2.3.20 Telephone Manufacturing Co

Telephone Manufacturing Co Ltd (TMC) was primarily a manufacture of telephones and associated equipment but also made electric master clocks and subsidiary dials (Pook 2014). At the outbreak of the First World War it was no longer possible to import telephones from Germany into the UK so TMC was formed to manufacture them. TMC went public as the Telephone Manufacturing Company (1920) Ltd. In 1924 it was known as the Telephone Manufacturing Company Limited and, after a re-organisation of activities, as the Telephone Manufacturing Co. 1929 Ltd. TMC's founder, F T Jackson, died on 14 August 1959. In the 1960s, TMC was taken over by the Pye Group, and became Pye TMC Ltd.

TMC's patent for synchronous clocks was awarded in February 1932. The date of the trademark Temco is usually given as ca 1935, but it was probably in use before February 1932. Production of Temco synchronous clocks appears to have started in 1931 and ended sometime after about 1952. They used their own synchronous clock movements and sometimes used synchronous clock movements made by another manufacturer.

2.3.21 *Westclox Ltd*

Westclox Ltd had its origin as the United Clock Company, formed by Charles Stahlberg and others, on 5 December 1885 in Peru, Illinois, USA (Wikipedia 2014). The intention was to manufacture clocks based on a patented technological innovation by Stahlberg. The company went bankrupt and in 1887 was reorganised as the Western Clock Company. It again went bankrupt and in 1888 was reorganised by F. W. Matthiessen as the Western Clock Manufacturing Company, shortened to Western Clock Company in 1912. The company first brought the Big Ben alarm clock to market in 1909. The modern trademark of the company, Westclox, appeared on the backs of Big Ben alarm clocks from 1910 to 1917, and on Big Ben dials as early as 1911. The trademark was registered on 18 January 1916. The Western Clock Co Ltd. was incorporated in 1919, and in 1931 merged with Seth Thomas Clock Company, with both becoming divisions of General Time Corporation. The Westclox unit became known as Westclox Division of General Time Corporation in 1936.

Westclox started making synchronous electric clocks in La Salle, Illinois, USA in 1931 (Linz 2004; ClockDoc 2014; Grace's Guide 2014). The first clocks were fitted with Sangamo movements, made by Westclox under licence. They started using their own design of synchronous motors in 1932. Non war production stopped in 1942, and normal production resumed in 1946.

A British subsidiary, Westclox Ltd, started making synchronous electric clocks in Dumbarton, Scotland in 1948. Westclox had originally planned to start production in Scotland in 1939, but the Second World War intervened and it was not until 1948 that they were able to fully commission their factory in Dumbarton. The Westclox Ltd factory in Dumbarton was intended to be a full manufacturing plant; with only the basic raw materials being bought in from outside suppliers. The initial plan was to produce clocks and watches using parts produced by the La Salle, Illinois, USA and Peterborough, Canada factories. Hence, early synchronous clocks incorporated foreign parts so are not marked as British Made. Later clocks were marked Made in Scotland. Westclox Ltd were a successful synchronous clock manufacturer. However, by 1988 the future of Westclox Ltd in Scotland was almost over and the company closed in 2001. They used their own synchronous clock movements.

References

Anonymous (1932) Synchronous clock conference. Horol J 75(892):255

Anonymous (1934) Ingersoll limited. Horol J 77(916):101

Anonymous (1936) Synchronous clock conference. Agreement between synchronous electric clock manufacturers and approved wholesalers. Horol J 78(931):8

Anonymous (1937a) Sterling Clock Co. Ltd. Horol J 79(948):13

Anonymous (1937b) Sterling Clock Co. Ltd. Horol J 80(949):11

Anonymous (1937c) British synchronous clock conference. Horol J 79(943):14

Anonymous (1937d) Testing synchronous clocks. Horol J 80(949):7

Anonymous (1937e) 'Ismay' electric clocks. Horol J 79(943):35

Anonymous (1938) Sterling Clock Co Ltd. Horol J 80(959):15

Anonymous (1939a) Current domestic clock design. Horol J 81(970):186–195

Anonymous (1939b) Britain's newest clock factory. Horol J 81(974):397–422

Anonymous (1940a) Ten years of progress in making and selling synchronous electric clocks. Horol J 82(984):1–3

Anonymous (1940b) Garrard timepiece. British Clock Manufacturer. Suppl Horol J 82(984):26

Anonymous (1945a) A review of the industry. Horol J 87(1046):361–363

Anonymous (1945b) Production news. Horol J 87(1044):287

Anonymous (1947a) Scottish electric clocks. Horol J 89(1066):344

Anonymous (1947b) Horological section of the B.I.F. A brief summary of the main exhibits. Horol J 89(1064):233–237

Anonymous (1947c) Elco clocks & watches. Horol J 89(1069):532

Anonymous (1949) Times goes quickly by road. Commercial motor, 16 Dec 1949:42

Anonymous (1951) Few clocks and watches. Horol J 93(1108):26

Anonymous (1962) BS 472: 1962 mains-operated synchronous clocks. British Standards Institution, London

Anonymous (2003) BS EN 60335-2-26:2003+A1:2008 household and similar electrical appliances. Safety particular requirements for clocks. British Standards Institution, London

Barrett DW (1946) Big future for British clocks. Official report of B.H.I. lecture. Horol J 88(1050):104–106

Bird C (2003) Metamec. The clockmaker. Antiquarian Horological Society, Dereham

ClockDoc – the Electric Clock Archive. http://www.electricclockarchive.org/. Accessed 2014

eHow (2014) http://www.ehow.co.uk/info_12269131_wages-1930s.html. Accessed 2014

Grace's Guide http://www.gracesguide.co.uk/. Accessed 2014

Li TD (1997) New Haven clocks and watches. Arlington Book Co, Fairfax

Lines MA (2012) Ferranti synchronous electric clocks. Paperback edition with corrections. Zazzo Media, Milton Keynes

Linz J (2001) Electrifying time: Telechron and GE clocks. Schiffer Publishing Co., Atglen

Linz J (2004) Westclox electric. Schiffer Publishing Co., Atglen

Miles RHA (2011) Synchronome. Masters of electrical time keeping. The Antiquarian Horological Society, Ticehurst

Nye J (2014) A long time in making. The history of Smiths. Oxford University Press, Oxford

Pook LP (2014) Temco Art Deco domestic synchronous clocks. Watch Clock Bull 56/1(407):47–58

Seager JH (1937) British clock cases. Horol J 80(949):70–72

Smith AG (1937) The re-establishment of the British clock making industry. Horol J 80(949):30, 32, 34, 36

Smith B (2008) Smiths domestic clocks, 2nd edn. Pierhead Publications Limited, Herne Bay

Wikipedia http://en.wikipedia.org. Accessed 2014

Chapter 3
How a Synchronous Clock Works

Abstract A synchronous electric clock is driven by a synchronous motor, and how this works is not obvious. The speed at which a synchronous motor rotates is determined precisely by its design and the frequency of the AC supply, hence the use of the adjective synchronous. If a running synchronous movement is examined all that can be seen is a rotating rotor which drives the hands through the reduction gear. In British made synchronous clocks the motor often, but not always, rotates at 200 rpm. Electric motors, including synchronous electric motors, are driven by magnetic forces produced by interacting magnetic fields. These magnetic fields are invisible. By contrast, a mechanical clock has an escapement which can be seen operating. How a synchronous motor works is described in detail, including the magnetic impulses which drive the motor. A synchronous motor is not inherently self starting. In a non self starting clock there is a manually operated device which runs the motor up to the synchronous speed in the correct direction, whereas in a self starting clock this is done automatically. Self starting methods rely on manipulation of magnetic fields, and two methods used in synchronous clocks are described. Designs of reduction gear used in synchronous movements vary widely. The striking and chiming work used in some synchronous clocks is described. The operation of a synchronous time switch is described.

3.1 Introduction

A synchronous electric clock is driven by a synchronous motor, and how this works is not obvious. If a running synchronous movement is examined all that can be seen is a rotating rotor which drives the hands through the reduction gear. In British made synchronous clocks the rotor often, but not always, rotates at 200 rpm. Electric motors, including synchronous electric motors, are driven by magnetic forces produced by interacting magnetic fields. These magnetic fields are invisible. By contrast, a mechanical clock has an escapement which can be seen operating (Pook 2011).

A magnet is a material or object that produces a magnetic field (Davidge and Hutchinson 1909; Walker 2011; Wikipedia 2014). This magnetic field is responsible for the most notable property of a magnet: a magnetic force that attracts or repels other magnets. The strength of a magnetic field at any point is the magnetic flux.

© Springer International Publishing Switzerland 2015   33
L.P. Pook, *British Domestic Synchronous Clocks 1930–1980*, History
of Mechanism and Machine Science 29, DOI 10.1007/978-3-319-14388-0_3

Fig. 3.1 Permanent bar
magnet

Fig. 3.1 Permanent bar
magnet

The magnetic reluctance, or magnetic resistance, of a material is a measure of its resistance to a magnetic field. It is analogous to a material's electrical resistance. A magnetic flux follows the path of least magnetic reluctance is just the same way that an electric current follows the path of least electrical resistance. In a magnet the magnetism which produces the magnetic flux is concentrated at what are called magnetic poles.

There are two types of magnet. A permanent magnet has its own persistent magnetic field and is made from what is called a magnetically hard material. A permanent bar magnet has a magnetic pole at each end. A permanent bar magnet of a type used for teaching purposes is shown in Fig. 3.1. As is conventional, the poles are coloured red and black, and marked N (North) and S (South). This notation derives from the use of a pivoted bar magnet in a magnetic compass. The North pole points to the Earth's magnetic North and the South pole to the Earth's magnetic South. A North pole or a South pole cannot exist in isolation. A temporary magnet is made from what is called a magnetically soft material. It is temporarily magnetised by placing it within a magnetic field. This is called magnetic induction. When this magnetic field is removed the induced magnetism disappears. In an electromagnet the temporary magnet is produced by a coil through which an electric current is passed. The overall strength of a magnet is measured by its total magnetic flux. The extent to which a material retains magnetism when it is removed from a magnetic field is called its coercivity. Hard magnetic materials have high coercivity whereas soft magnetic materials have low coercivity.

If two magnets are brought together a North pole attracts a South pole, but two North poles repel each other, as do two South poles. In an ideal situation the force between two poles is inversely proportional to the square of the distance between them (Davidge and Hutchinson 1909; Stott 1946). When an item, such as a nail, made from a magnetically soft material is in the vicinity of a permanent magnet, magnetism of appropriate polarity is induced in the nail and it is attracted to the permanent magnet. When an item made from a magnetically hard material is within the vicinity of a permanent magnet it becomes magnetised. This can be a nuisance, for example when the hairspring of watch becomes magnetised and its timekeeping is affected. Care is therefore needed when handling magnets.

Much effort has been devoted to the development of materials for magnetic purposes since the magnetic design of devices such as electric motors is as important as their mechanical design (Wikipedia 2014). One aspect of magnetic design is that unwanted eddy currents induced by moving magnetic fields need to be kept to a minimum. Synchronous clocks usually use soft iron when a soft magnetic material is required. Early development of synchronous clocks was helped by the development of cobalt steel as a hard magnetic material. Some later clocks used ferrite. A ferrite is a ceramic made from iron oxides such as haematite (Fe_2O_3) or magnetite (Fe_3O_4) as well as oxides of other metals. Ferrites are, like most other ceramics, hard and brittle. For magnetic applications ferrites have the advantage of being non conducting so there are no eddy currents.

Discussion of synchronous motors used in synchronous clocks in this chapter is intended to assist understanding, and is largely qualitative. In particular, detailed discussion of the magnetic design of the synchronous motors is outside its scope. For descriptions of individual synchronous clock movements see Chap. 11.

3.2 Synchronous Motors

An electric motor consists of a rotor which rotates within a stator. It is driven by the magnetic fields associated with the rotor and stator in the same way that the magnetic pointer of a compass aligns itself with the Earth's magnetic field. Motors can be designed to use either DC (direct current) or AC (alternating current).

The time keeping of a synchronous clock depends on a synchronous motor, which is powered by an alternating current (AC) supply of electricity. The speed at which a synchronous motor rotates is determined precisely by its design and the frequency of the AC supply, hence the use of the adjective synchronous. In the UK the frequency is 50 Hz (cycles per second). The detail design of synchronous motors used in synchronous clocks varies widely. General descriptions have been given by a number of authors, including Britten (1978), Miles (2011), Philpott (1935), Robinson (1942) and Stott (1946). The descriptions in Sects. 3.2.1, 3.2.2, 3.2.3, and 3.2.4 are intended to aid understanding of synchronous motors, but not to provide detail information on how they are designed.

A synchronous electric motor is an AC motor in which the stator is an electromagnet. Poles (projections) on the stator become magnetic poles which are alternately N and S as the AC current changes direction. Magnetic poles (projections) on the rotor are attracted in synchronism by the magnetic poles. At synchronism the rotor is said to be phase locked to the alternating field magnetic field produced by the stator. In a similar way the Moon always has the same face towards the Earth. This is because the rotation of the Moon about its axis is phase locked to its revolution about the Earth.

Synchronous electric motors are of two basic types. In a reluctance motor (Sect. 3.2.1), sometimes called an attraction motor, the rotor is made from a magnetically soft material, usually soft iron, and is magnetised in alternate directions by the alternating magnetic field produced by the stator. In a magnetised motor (Sect. 3.2.2) the rotor is a permanent magnet made from a magnetically hard material. Sensitivity to voltage drops varies widely (Moore 1936). The power is so small that there is little trouble due to overheating in synchronous clocks (Philpott 1934a).

A basic synchronous motor with the power turned on has three stable states: stationary, or running in either direction at the synchronous speed. A starting method is therefore needed to run the motor up to the synchronous speed in the correct direction. Basic synchronous motors can be modified to be self starting, and there are several methods of doing this (Sect. 3.2.4). Of the movements described in Chap. 11 approximately equal numbers have reluctance and magnetised motors.

The number of poles on the rotor and stator, and their spacing, depends on the detail design. In a reluctance motor the poles on the rotor are equally spaced and the number of poles is twice an odd number. There is usually a smaller number of poles on the stator. The number and their spacing is determined by the need to match their polarity with the polarity of poles on the stator. The situation is reversed in a magnetised motor. The poles on the stator are equally spaced, the number of poles is twice an odd number, and there is usually a smaller number of poles on the rotor.

Care needs to be taken to ensure that the rotor bearings are durable and adequately lubricated (Wise 1951). Felt pads or sintered bearings are often used to provide an oil reservoir.

3.2.1 Reluctance Synchronous Motors

A reluctance synchronous motor is synchronous motor in which the rotor is a temporary magnet made from a magnetically soft material. There are some differences in detail but most reluctance synchronous motors used in clocks have the same main features. An exception is the Marigold movement which is described in Sect. 11.10. The main features of a typical reluctance synchronous motor are illustrated in Fig. 3.2. This shows a Metamec Type 1 movement (Sect. 11.12.1). The laminated U-form stator (Fig. 3.2b) is made from a magnetically soft material. It is an electromagnetic with a coil at its apex, and it is laminated to reduce unwanted eddy currents. The poles on the left hand side and right hand side of stator are of opposite polarity. The rotor (Fig. 3.2a, c) is also made from a magnetically soft material. It has 30 poles and rotates at 200 rpm. The stator has a smaller number of poles, but the angular spacing is the same.

The voltage of an AC supply varies sinusoidally with time about zero as shown in Fig. 3.3. Its average value is zero. The nominal voltage of an AC supply is its rms (root mean square) value. This is calculated by squaring instantaneous values, then finding the mean (average) and taking the positive square root. Since squares are always positive the rms has a positive value. The magnetic field of a stator induces magnetism of opposite polarity in an adjacent rotor pole. Hence, the poles always attract each other irrespective of the sign of the voltage. Hence, the resulting magnetic impulses, for a 50 Hz AC supply, have a frequency of 100 Hz. This is shown schematically in Fig. 3.4. The rotor is phase locked so it takes 0.01 s for a rotor to move from opposite one stator pole to opposite the next stator pole. It follows immediately that a rotor with 30 poles rotates at 200 rpm.

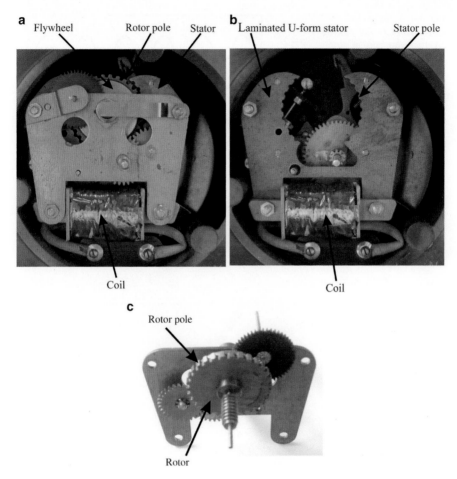

a Flywheel Rotor pole Stator **b** Laminated U-form stator Stator pole

Coil Coil

c Rotor pole

Rotor

Fig. 3.2 Metamec Type 1 movement, (**a**) Front view, (**b**) Front view with the front plate and rotor assembly removed, (**c**) Rotor assembly

Fig. 3.3 Variation of voltage with time for an alternating current supply for a frequency of 50 Hz

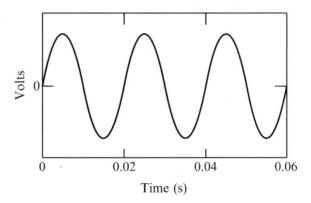

Volts

0

0   0.02   0.04   0.06

Time (s)

Fig. 3.4 Variation of
magnetic force with time for
an alternating current supply
with a frequency of 50 Hz

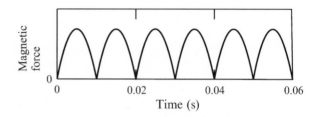

Fig. 3.5 Rotor assembly
from a Garrard movement

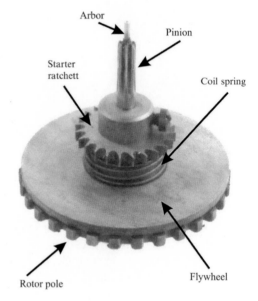

In a reluctance motor, a flywheel (Figs. 3.2a and 3.5) is usually added to smooth
out the rotation of the rotor (Stott 1946). The rotor is fixed to the arbor. The flywheel
is free to rotate and is loosely linked to the rotor. A typical arrangement is shown
in Fig. 3.5 which shows the rotor assembly from a Garrard Early movement (Sect.
11.7.1). A coil spring is connected to the flywheel at one end and to a starter ratchet
at the other end. A pinion fixed to the arbor drives the reduction gear. The starter
ratchet is fixed to the arbor. This is part of an indirect manual starter (Sect. 3.2.4.2).

3.2.2 Magnetised Synchronous Motors

As the name of a magnetised synchronous motor implies, the rotor is permanently
magnetised. Hence, the rotor poles do not change polarity as the AC voltage on the
stator coil changes sign. This means that the stator poles have to be of alternate
polarity. Typically, this is achieved by using a stator with a full number of poles
which are interleaved. The rotor usually has a smaller number of poles. Care has

Fig. 3.6 Rear view of a Temco Mark V movement with the back of the motor cover removed

Stator with interleaved poles

Rotor

Coil

Leads to coil

to be taken to ensure that these are appropriately spaced. The drive to the rotor is smoother than in a reluctance motor so flywheels are not used. Main features of magnetised motors vary so much that none can be described as typical. Some of the arrangements that have been used are described in this section.

Temco Mark V Movement Figure 3.6 shows a rear view of a Temco Mark V movement (Sect. 11.15.4) with the back of the motor cover removed, The permanently magnetised rotor has a pair of poles at each end. The pairs of poles at each end of the rotor are of opposite polarity. The stator is an electromagnet which consists of two discs made from magnetically soft material. The coil is sandwiched between the two discs so they are of opposite polarity. Each disc has 15 poles which are bent at right angles such that they are interleaved and separated by air gaps. The total of 30 poles on the stator means that the rotor rotates at 200 rpm.

Magneta Later Movement A rear view of a Magneta Later movement Sect. (11.8.3) with the back plate removed is shown in Fig. 3.7. The permanently magnetised rotor has six pairs of poles. The pairs are of alternating N and S polarity. The overlapping laminated stator is made from a magnetically soft material. It is an electromagnetic with a coil at its apex, and it is laminated to reduce unwanted eddy currents. The ends overlap and are of opposite polarity. Each end has a hole with 15 poles, which are bent at right angles so that they are interleaved and separated by air gaps. The total of 30 poles on the stator means that the rotor rotates at 200 rpm.

Goblin M.6. Self Starting Movement An angled front view of a Goblin M.6. self starting movement (Sect. 11.8.1) is shown in Fig. 3.8. The permanently magnetised rotor has six pairs of poles. The pairs of poles are of alternating N and S polarity. The doughnut shaped stator is an electromagnetic which consists of two parts made

Fig. 3.7 Rear view of a
Magneta Later movement,
with the back plate removed

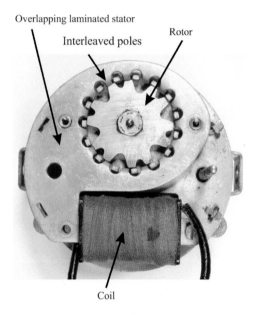

Overlapping laminated stator

Interleaved poles

Rotor

Coil

Fig. 3.8 Angled front view
of Goblin M.6. self starting
movement

Stator

Rotor

Interleaved poles

Leads to coil

from a magnetically soft material. The coil is sandwiched between the two parts so
they are of opposite polarity. Each part has 15 poles which are bent at right angles
such that they are interleaved and separated by air gaps. The total of 30 poles on the
stator means that the rotor rotates at 200 rpm. The movement has an impulse self
start (Sect. 3.2.4.3).

Fig. 3.9 ¾ top view of Smith
QGEM movement

Rotor

Coil   Stator with interleaved poles

Smith QGEM movement   A three-fourths top view of a Smith QGEM movement
(Sect. 11.13.8) is shown in Fig. 3.9. The permanently magnetised ferrite disc
rotor has no visible poles. Magnetically it has six pairs of poles. The pairs are of
alternately N and S polarity. The stator is an electromagnet which consists of two
discs made from magnetically soft material. The coil is sandwiched between them
so they are of opposite polarity. Each disc has 15 poles which are bent at right angles
so that they are interleaved and separated by air gaps. The total of 30 poles on the
stator means that the rotor rotates at 200 rpm.

Westclox Type No. BM25 Bottom Set Movement   Figure 3.10 shows a front
view of a Westclox Type No. BM25 bottom set movement (Sect. 11.16) with the
front plate removed. The laminated U-form stator is made from a magnetically
soft material. It is an electromagnetic with a coil at its apex, and it is laminated
to reduce unwanted eddy currents. The poles on the left hand side and right hand
side of stator are of opposite polarity. The rotor consists of two discs made from
a magnetically soft material, with a magnet sandwiched between them so that the
discs are of opposite polarity. Each disc has 15 poles which are bent at right angles
such that they are interleaved and separated by air gaps. The total of 30 poles means
that the rotor rotates at 200 rpm. The stator has a smaller number of poles.

3.2.3   Magnetic Impulses

The magnitude and form of the sequence of identical magnetic impulses which drive
a synchronous motor depend on the phase. This is the relationship between when a

Fig. 3.10 Front view of a
Westclox Type No. BM25
bottom set movement with
front plate removed

Laminated U-form rotor

Rotor

Interleaved poles

Coil

rotor pole is opposite a stator pole and when the AC current has its maximum value. The phase lag is the time that elapses between when a rotor pole is opposite a stator pole and when the AC current has its maximum value.

A series of numerical calculations were carried out using a spreadsheet to determine the form and magnitude of magnetic impulses in reluctance motors and in magnetised motors. Selected results are shown in Sects. 3.2.3.1 and 3.2.3.2 to illustrate characteristics of reluctance and magnetised motors. As a simplification magnetic poles were assumed to be points. This means that the magnetic force between two magnetic poles is inversely proportional to the square of the distance between them (Sect. 3.1). As another simplification only the magnetic forces between a rotor pole and the two adjacent stator poles were calculated. These are referred to as the first and second stator poles. The number of poles was taken as 30, and the AC current frequency was taken as 50 Hz. Measurements on a number of synchronous motors in clocks of various makes showed that the clearance between the rotor poles and the stator poles is approximately 100th of the rotor diameter, measured over the poles, so this clearance was used. The magnetic forces between rotor poles and stator poles can be resolved into two components. One component is tangential and was used in calculations. The tangential forces produce the moments about the axis of the rotor and drive it. The other component is radial so has no effect.

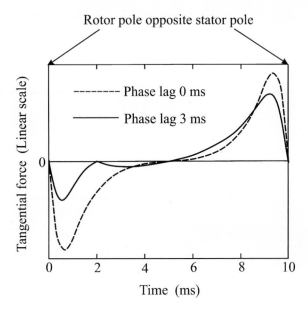

Fig. 3.11 Tangential forces in a reluctance motor, 50 Hz

3.2.3.1 Reluctance Motors

Figure 3.11 shows the tangential forces produced in a reluctance motor, by a 50 Hz AC supply, for two different values of phase lag. A curve repeats exactly so there is a sequence of identical magnetic impulses which drives the rotor. For phase lag 0 ms and time 0 ms, the rotor pole is opposite the first stator pole, and the tangential force is zero. As the rotor pole moves away from the first stator pole the rotor pole is attracted to the first stator pole, and the tangential force is negative. It decreases to a minimum, increases to zero, when the rotor pole is midway between the two stator poles, and the rotor pole is equally attracted to both. The tangential force then increases to a maximum, and decreases to zero, when the rotor pole is opposite the second stator pole. The curve has rotational symmetry.

For phase lag 3 ms and time 0 ms, the rotor pole is opposite the first stator pole and the tangential force is zero. As the rotor pole moves away from the first stator pole the rotor pole is attracted to the first stator pole and the tangential force is negative. It decreases to a minimum, increases to zero, then decreases to another minimum. The tangential force next increases to zero when the rotor pole is midway between the two stator poles, and the rotor pole is equally attracted to both. Finally, the tangential force increases to a maximum, and decreases to zero, when the rotor pole is opposite the second stator pole. The magnetic impulse is in three parts. The first two are negative and try to drive the rotor backwards, the third is positive and drives the rotor forward. This characteristic is the reason why flywheels are needed in reluctance motors.

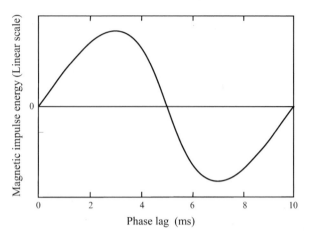

Fig. 3.12 Magnetic impulse energies in a reluctance motor, 50 Hz

The total magnetic impulse energy is given by the area under a tangential force curve, where the area below zero is taken as negative. For a 50 Hz AC supply the phase lag can vary from 0 to 10 ms, and values of the magnetic impulse energy for this range are shown in Fig. 3.12. The curve has rotational symmetry. This reflects the fact that a synchronous motor can run equally well in either direction. The magnetic impulse energy has a maximum at a phase lag of 3 ms and a minimum at 7 ms. These are not sharply defined. Once a reluctance motor is running in synchronism (phase locked) it is stable in the sense that it will continue to run at the synchronous speed provided that the torque produced by the sum of the magnetic impulses due to all rotor pole/stator pole pairs is sufficient to overcome the load on the motor. If the load becomes too great the motor usually drops out of synchronism and stops.

Experimental data were obtained by Anonymous (1948) who used a set up in which the stator was represented by a pair of electromagnets energised by a DC supply, and the rotor by a pivoted magnet. The force on the magnet was measured with the magnet in a range of positions. The results obtained were broadly similar to those shown in Figs. 3.11 and 3.12. In particular the structure of magnetic impulses is confirmed. The maximum magnetic impulse is at a phase lag of 3.1 ms: nearly the same as in Fig. 3.12.

3.2.3.2   Magnetised Motors

Figure 3.13 shows tangential forces produced in a magnetised motor, by a 50 Hz AC supply, for two different values of phase lag. A curve repeats exactly so there is a sequence of identical magnetic impulses which drives the rotor. For phase lag 0 ms and time 0 ms the rotor pole is opposite the first stator pole, and the tangential force is theoretically zero. In the figure it has a small positive value. This is an artefact of the simplifications used. As the rotor pole moves away from the first stator pole the

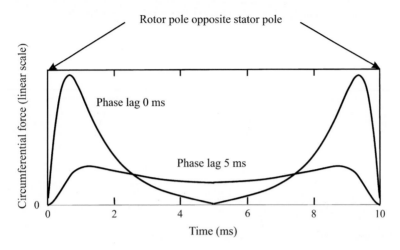

Fig. 3.13 Tangential forces in a magnetised motor, 50 Hz AC

rotor pole is repelled by the first stator pole, and the tangential force is positive. It increase to a maximum, and then decreases to zero, when the rotor pole is midway between the two stator poles. It then increases to a maximum, and decreases to zero, when the rotor pole is opposite the second stator pole. The curve is symmetrical. The characteristic absence of negative values of the tangential force is the reason why flywheels are not needed in magnetised motors.

For phase lag 5 ms and time 0 ms, the rotor pole is opposite the first stator pole, and the tangential force is zero. As the rotor pole moves away from the first stator pole the rotor pole is repelled by the first stator pole, and the tangential force is positive. It increase to a maximum, and then decreases to a minimum, when the rotor pole is midway between the two stator poles. It then increases to a maximum, and decreases to zero, when the rotor pole is opposite the second stator pole. The curve is symmetrical. The tangential force is never negative. This characteristic is the reason why flywheels are not needed in magnetised motors.

The total magnetic impulse energy is given by the area under a tangential force curve. For a 50 Hz AC supply the phase lag can vary from 0 to 10 ms, and values of the magnetic impulse energy for this range are shown in Fig. 3.14. The curve is symmetrical. This reflects the fact that a synchronous motor can run equally well in either direction. The magnetic impulse energy has a maximum at a phase lag of 0 ms, and a minimum at 5 ms. These are not sharply defined. Once a magnetised motor is running it is stable in the sense that it will continue to run provided that the torque produced by the sum of the magnetic impulses due to all rotor pole/stator pole pairs is sufficient to overcome the load on the motor. A magnetised motor is not quite stable when it is in synchronism (phase locked). The shape of the magnetic impulse energy curve means that is possible for a magnetised motor to run at just below the synchronous speed, and experience shows that this occasionally happens.

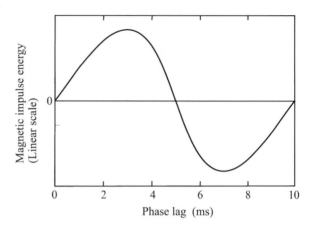

Fig. 3.14 Magnetic impulse energy produced in a magnetised motor by a 50 Hz AC current in the stator coil

3.2.4   Starting Methods

The synchronous motors used in synchronous clocks can be either non self starting or self starting. In a non self starting clock there is a manually operated device which runs the motor up to the synchronous speed in the correct direction, whereas in a self starting clock this is done automatically. If there is an interruption in the mains supply a clock stops and it is obvious that it is not showing the correct time. However, following an interruption a self starting clock re-starts, but it is not obvious that it is not showing the correct time. Some American made synchronous clocks have an outage indicator which shows that there has been an interruption, but there do not appear to be any British made synchronous clocks with this feature. Self starting methods rely on manipulation of magnetic fields and are usually used in synchronous clocks with magnetised motors. Three starting methods used on British synchronous clocks are described in Sects. 3.2.4.1, 3.2.4.2, and 3.2.4.3. Unreliability of starters used in British synchronous clocks was, and still is, a significant problem for users. A work around that can be used with some clocks is to remove the movement cover and give an appropriate wheel a push in the right direction. This needs to be done with care since live parts may be exposed.

3.2.4.1   Direct Manual Start

In the direct manual starting method a starter knob is turned manually in the direction indicated. Figure 3.15 is an example which shows the starter knob of a Ferranti Model No. 5 synchronous bedside clock (Sect. 8.4). The starter knob continues to rotate when the clock is running, and thus acts as a tell tale. This method is simple and reliable, but does need some skill, and the mischievous can start clocks backwards. There is a separate hand set knob (Sect. 3.2.5).

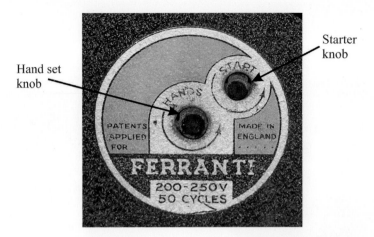

Starter
knob

Hand set
knob

Fig. 3.15 Starter and hand set knobs of a Ferranti Model No. 5 synchronous mantel clock

3.2.4.2 Indirect Manual Start

Indirect starting methods vary considerably in detail but all work on the same principle: a spring is tensioned and then released. When released the spring delivers an impulse of the correct magnitude to start the clock. The impulse runs the rotor up to a speed greater than the synchronous speed then, as it slows down, it drops into synchronism (Stott 1946). As an example, Fig. 3.16 shows a Clyde movement (Sect. 11.3). When the pivoted starter lever is moved to the left the spring (Fig. 3.16b) is tensioned, and the attached prong engages with rotor poles. When the lever is released the prong spins the rotor to the right and the clock starts. This method is reliable when the clock is new but thickening oil, wear and damage can mean that an appropriate impulse is not delivered, and the clock becomes difficult or impossible to start. Sometimes a prong or pawl engages with a starter ratchet, for example Fig. 3.5.

Sometimes there is a combined starter and hand set knob. The knob is pushed in to set the hands and when released the clock starts. This makes to difficult to set the time precisely.

3.2.4.3 Impulse Self Start

There are several self starting systems that have been used in magnetised synchronous clock motors (Holmes and Grundy 1935; Miles 2011; Wise 1951). The best description of how these work is given by Wise. One system is the impulse self start. The angular spacing of the poles of a synchronous motor is normally the same for all the poles. In an impulse self start synchronous motor the angular

Fig. 3.16 Clyde synchronous mantel clock, (**a**) Starter lever, (**b**) Top view of movement

spacing of the rotor poles is slightly uneven, and the magnetic field is out of balance as soon as the stator is energised. The amount of magnetic out of balance is predetermined by design. The resulting out of balance magnetic impulses make the rotor oscillate. The magnitude of these oscillations increases rapidly and the rotor drops into synchronisation. The motor is equally likely to run in either direction. A very lightly loaded pawl engaging with a single ratchet tooth is the usual method of preventing rotation in the wrong direction (Stott 1946). The Goblin M.6. self starting movement (Fig. 3.8) has an impulse self start. Although not obvious from the figure the rotor pole spacing is slightly uneven. The Magneta Later movement uses the same system (Sect. 11.8.3).

3.2.4.4   Shaded Pole Self Start

The shaded pole self starting method has the advantage that the motor always starts in the correct direction. No mechanical device is needed to prevent rotation in the wrong direction. Initially, the shaded pole self could not be used by most British manufacturers because its use was protected by an American owned patent. There do not appear to be any British made synchronous clocks with a shaded pole self start. The example shown in Fig. 3.17 is the movement of a President synchronous novelty clock. It is not British made. The shaded poles are created by two asymmetrically place copper coils surrounding the stator. Induced currents in these coils produce an out of balance magnetic field. The resulting out of balance magnetic impulses start the rotor in the correct direction.

Fig. 3.17 Movement of a
President synchronous
novelty clock

Coil Copper coils

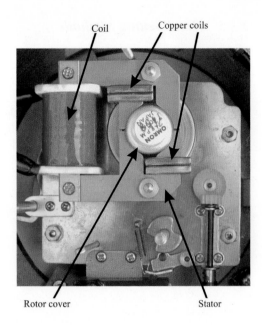

Rotor cover Stator

3.2.4.5 Synchronous Clocks Running Backwards

Occasionally, a self starting synchronous clock runs backwards following a power outage. This happens when the mechanism that ensures rotation in the correct direction fails to operate. In practice this is rare but it can have serious consequences, for example if a synchronous time switch is being used to control a domestic central heating system. This has been known to happen with a synchronous time switch based on a Smith Bijou movement (Sect. 11.13.5).

Mains distribution systems include circuit breakers to protect the system in the event of an electrical overload. These automatically reconnect to check whether the overload was temporary. A lightning strike can make this happen and produce a temporary power outage If a synchronous clock then runs backwards it gives the impression that a lightning strike has affected the clock directly. One such episode was reported fancifully by Anonymous (1947) as follows.

> Customers who were enjoying their beer in the bar of the Freshwater Bay Hotel, Isle of Wight, one evening recently, noticed with delight that instead of the hands of the clocks moving steadily towards closing time, they were going in the opposite direction! As the news spread, the bar filled up, and it seemed that a good time would be enjoyed by all for an indefinite period! By 10 o'clock the hands pointed to six, and the happy atmosphere in the bar improved every minute. All good thongs must come to an end, however, and the management, fearing that the powers behind the local licensing laws would not sympathise with their clocks, called 'time' and broke up the party.

A similar story is told by Read (2012) but with only one clock affected, which is more likely.

3.2.5   Hand Setting Methods

In domestic synchronous clocks with unglazed dials, or glazed dials with an opening door, the hands can be set to time by manually turning the minute hand. A friction drive allows this to be done while the clock was running. Seconds hands are not adjustable. In most domestics synchronous clocks with glazed dials the hands are not accessible and setting is by a hand set knob (Fig. 3.15). This is pushed in against a spring, turned to set the hands and released. Pushing the knob in engages a gear which disengages when the knob is released. In some domestic synchronous clocks the hand set knob is combined with starter knob. The knob is pushed in and released to start the clock. This arrangement is unsatisfactory because the clock stops when the knob is pushed in to set the hands, and restarts when the knob is released. This makes it difficult to set a clock to time.

3.3   Reduction Gear

The reduction gear is the train of gears in a synchronous clock that reduces the speed of a synchronous motor to that of the hands. A synchronous clock motor typically rotates at 200 rpm which makes it a low speed motor in electrical engineering terms, but it is high speed when compared with the 'scape wheel of a mechanical mantel clock. This typically rotates at 2 rpm.

Design requirements for the reduction gear of a synchronous clock differ from those for the going train of a mechanical clock (Sect. 1.3.1). Firstly, unlike a mechanical clock, the reduction gear is lightly loaded throughout because only enough power to drive the hands needs to be transmitted so only light wheels are needed. The wheels in Heavy Motion Work Movements (Sect. 11.13.4), used in clocks with large hands, are still quite light. Secondly, a large reduction needs to be provided. Since power does not have to be transmitted to an escapement it is possible to use worm gears in the reduction gear. This is sometimes done to reduce the number of wheels in the reduction gear. Thirdly, a synchronous motor runs continuously so wheels in the reduction gear do not need to be crossed out to reduce inertia. This is sometimes done in early clocks where existing production facilities for crossed out wheels for mechanical clocks were used to produce wheels for synchronous clocks. Finally, noise from the reduction gear needs to be kept to a low level. The motor speed is comparable with wheels at the end of the striking train of a mechanical clock. The sort of whirring noise which is usual when a clock is striking is not acceptable in a synchronous clock.

Designs of reduction gear used in movements vary widely. Descriptions of reduction gear used in movement have been given by a number of authors, but with little or no indication of why a particular design was chosen. Production costs were always a concern. Plastic gears were introduced in the 1960s because they were cheaper than metal gears (Miles 2011; Smith 2008). In early clocks the first wheel is usually fibre to reduce noise (Philpott 1934b). Of the movements described in Chap. 11 approximately equal numbers have a double worm reduction gear and a reduction gear which is similar to the going train of a conventional mechanical clock. There are also two movements with a triple worm reduction gear, and one with a single worm reduction gear. Major manufacturers made both worm reduction gears and reduction gears which are similar to the going train of a conventional mechanical clock. Minor manufacturers who only produced synchronous clocks for a short period only used the latter.

The reduction gear in a Marigold movement is unusual. It is described in Sect. 11.10.

3.4 Striking and Chiming Work

The design of the striking and chiming work using in British domestic striking and chiming domestic synchronous clocks was based on well established designs used in mechanical clocks (Stott 1946). Some of these designs are described by Britten (1978). Striking or chiming work in a mechanical clock is usually powered by either a falling weight or a mainspring.

Smiths Industries were the major British manufacturer of striking and chiming synchronous clocks. They also supplied chiming synchronous movements to at least one other manufacturer (Sect. 7.28). Two methods were used in arranging drives to the striking and chiming work in synchronous clocks. In one method, used in early chiming clocks, the striking and chiming work is powered by mainsprings, as in mechanical clocks. The mainsprings have to be wound up when the clock is first used, and are subsequently rewound automatically by the synchronous motor. This method was used in the Smith Type 1 Chiming Movement (Sect. 11.13.11). This is the only movement to use the mainspring method. In the other method, a jockey wheel engages automatically, when needed, with a train from the synchronous motor, and disengages automatically when striking, or chiming and striking, is complete. This method was used in the Smith Type 1 Striking Movement (Sect. 11.13.9). This was the first movement to use the jockey wheel method. How the methods work is illustrated by descriptions of the Smith Type 1 Chiming Movement and the Smith Narrow Striking Movement (Sect. 11.13.10).

Smith Type 1 Chiming Movement This was the first Smith chiming movement The striking side of a Smith Type 1 chiming movement is shown in Fig. 3.18. Arrangements on the chiming side are similar. The train of gears that drives the striking work is called the striking train, and is driven by a mainspring enclosed in a barrel. When the clock is first used the mainspring has to be wound up by means of a key on the winding square, which is on the end of the mainspring arbor. The striking train is driven via a wheel on the mainspring arbor. The mainspring is rewound

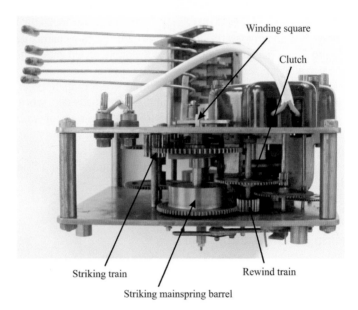

Fig. 3.18   Striking side of Smith Type 1 chiming movement

via the mainspring barrel by a train of gears, called the rewind train, which is permanently connected to the synchronous motor. The rewind train includes a spring loaded clutch which ensures that the drive is disconnected when the mainspring is fully rewound.

Smith Narrow Striking Movement   A top view of a Smith Narrow striking movement is shown in Fig. 3.19a. The striking train includes a jockey wheel. This is mounted on a pivoted arm and drops into engagement when needed, completing the striking train. This is a much simpler system than the automatic rewind system used in the Smith Type 1 Chiming Movement, and achieves the same effect. Figure 3.19b is a rear view of the front plate. This shows the jockey wheel arm and the two wheels with which the jockey wheel engages.

3.5   Synchronous Time Switches

The actual electrical switches used in synchronous time switches are mechanical switches operated by a cam that rotates once per 24 h. Lobes on the cam are adjustable by the user so that the time switch can be set to turn electrical devices on and off at desired times. A modern synchronous time switch for general domestic use can be plugged into a socket, and has a socket for the electrical device. These

Jockey wheel

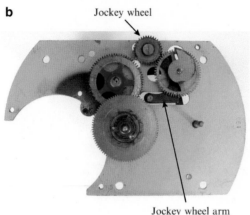

Jockey wheel arm

Fig. 3.19 Smith Narrow striking movement, (**a**) Top view, (**b**) Rear view of front plate

are now, confusingly, called mechanical time switches. This is to distinguish them from quartz time switches in which the actual switch is a solid state device. These are now called digital time switches.

As an example, a Wickes Model No WSEG general purpose synchronous time switch is shown in Fig. 3.20a. It was bought new in 2006 and is not British made. The output socket is opposite the input plug. Switching times are set by pushing segments down to form lobes on a cam, as shown in Fig. 3.20b. Segments forming the lobe have become discoloured during use of the time switch. The segments provide increments of 15 min. The 24 h dial rotates with the cam. A lobe on the cam pushes a lever to operate a microswitch (Fig. 3.20c). Very little power is needed so the synchronous motor is much smaller than those used in synchronous clocks. The coil, stator and rotor cover are visible in the figure. The movement has an impulse

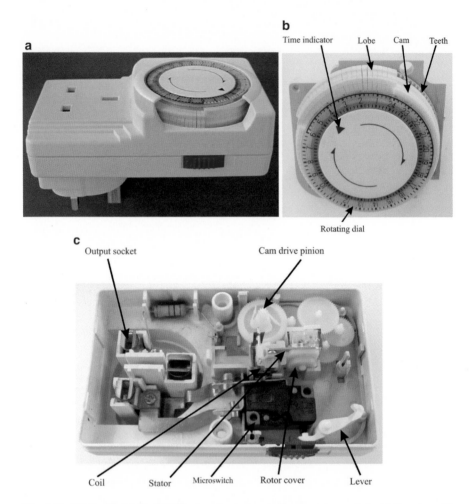

Fig. 3.20 Wickes Model No WSEG general purpose synchronous time switch, (**a**) Top view, (**b**) Cam assembly, (**c**) Top view with cover and cam assembly removed

self start (Sect. 3.2.4.3). The cam drive pinion is the final pinion in the single worm reduction gear. It drives a wheel which is integral with the cam. Some of the teeth of this wheel are visible in Fig. 3.20b.

References

Anonymous (1947) Lightning reverses clocks. Horological J 89(1069):517
Anonymous (1948) Manual-start synchronous motor. Horological J 90(1077):338–339
Britten FJ (1978) The watch & clock makers' handbook, dictionary and guide. 16th edn. Revised by Good R. Arco Publishing Company, New York

Davidge HT, Hutchinson RW (1909) Technical electricity, 2nd edn. University Tutorial Press Ltd., London

Holmes W, Grundy E (1935) Small self-starting synchronous time motors. Horological J 77(923):348, 350

Miles RHA (2011) Synchronome. Masters of electrical time keeping. The Antiquarian Horological Society, Ticehurst

Moore GE (1936) Miniature synchronous motors in service. Horological J 78(933):3–4

Philpott SF (1934a) Synchronous clocks – troubles and remedies. Horological J 77(914):55–56

Philpott SF (1934b) Synchronous clocks – troubles and remedies. Horological J 77(911):88, 90

Philpott SF (1935) Modern electric clocks, 2nd edn. Sir Isaac Pitman & Sons Ltd., London

Pook LP (2011) Understanding pendulums. A brief introduction. Springer, Dordrecht

Read D (2012) The synchronous mains revolution and 'Time gentlemen, please'. Antiquarian Horology 33(4):487–492

Robinson TR (1942) Modern clocks. Their repair and maintenance, 2nd edn. N A G Press Ltd., London

Smith B (2008) Smiths domestic clocks, 2nd edn. Pierhead Publications Limited, Herne Bay

Stott HT (1946) Electricity and horology. Horological J 88(1048):23–30

Walker J (2011) Halliday & Resner. Principles of physics. International student version, 9th edn. Wiley, Hoboken

Wikipedia (2014) http://en.wikipedia.org. Accessed 2014

Wise SJ (1951) Electric clocks, 2nd edn. Heywood & Company Ltd., London

Chapter 4
Synchronous Clock Cases

Abstract Clocks for domestic use have been produced for four centuries. Historically, clock case design styles have been based on furniture design styles with the obvious constraints that the case must be big enough to enclose the movement, the dial must be legible, it must be possible to set the hands, and only available materials, usually wood, could be used. Cases were usually not much bigger than needed to enclose the movement. Addition requirements for synchronous movements are: provision for the mains lead and, for non self starting movements, access to a starting knob or lever. An adjacent socket (AC outlet) is needed. If this is behind the clock then the mains lead can be concealed. Contemporary advertisements always illustrated domestic synchronous clocks without any visible wiring. A wide range of materials was used for synchronous clock cases. In the 1930s materials no longer limited, but assisted case designers. Materials used included solid wood, plywood, brass, pewter, glass, porcelain, stone, and Bakelite. Later, aluminium and Perspex were also used. Domestic synchronous clock cases were of four main types: mantel clocks, bedside clocks, wall clocks, and granddaughter clocks. Some of the artistic styles used for domestic synchronous clock cases are illustrated.

4.1 Introduction

Clocks for domestic use have been produced for four centuries. Historically, clock case design styles have been based on furniture design styles (Seager 1936a, b, 1937) with the obvious constraints that the case must be big enough to enclose the movement, the dial must be legible, it must be possible to set the hands, and only available materials, usually wood, could be used. Cases were usually not much bigger than needed to enclose the movement. Addition requirements for synchronous movements are provision for the mains lead and, for non self starting movements, access to a starting knob or lever. An adjacent socket (AC outlet) is needed. If this is behind the clock then the mains lead can be concealed. Low profile plugs and sockets (Fig. 1.6) used to be available for this purpose, and were widely used for wall clocks. Some wall clocks needed a purpose made recess to accommodate a protrusion. Contemporary advertisements always illustrated domestic synchronous clocks without any visible wiring.

© Springer International Publishing Switzerland 2015 57
L.P. Pook, *British Domestic Synchronous Clocks 1930–1980*, History
of Mechanism and Machine Science 29, DOI 10.1007/978-3-319-14388-0_4

Available space for a clock is sometimes an important constraint in clock case design. In the 1930s houses were being built with much narrower mantelpieces so shallow case clocks were needed. Shallow case timepiece synchronous mantel clocks were being advertised in 1935 (Anonymous 1935) and shallow case striking and chiming synchronous mantel clocks in 1936 (Anonymous 1936). A wide range of materials was used for synchronous clock cases (Anonymous 1939). In the 1930s materials no longer limited, but assisted case designers. Materials used included solid wood, plywood, brass, pewter, glass, porcelain, stone, and Bakelite. Later, aluminium and Perspex were also used. Cases were usually mass produced.

In synchronous clocks the traditional link between the form of a clock movement and the form of its case was lost. In other words, form no longer followed function. Cases were sometimes much larger than needed to enclose the movement. A feature of some synchronous clock cases is that the movement cover is part of the case, and sometimes projects behind the rest of the case. The names of synchronous clock case designers are usually unknown, an exception is W N Duffy who worked for Ferranti (Lines 2012).

Domestic synchronous clocks in the galleries in Chaps. 7, 8, 9, and 10 are clocks that were available for examination, and are partly a personal choice. The criteria for inclusion are that a clock, including its movement, is in reasonably original condition, was made in the UK, and was intended for use on the UK 50 Hz mains supply. They are believed to be broadly representative of the thousands of domestic synchronous clock case designs that were produced.

4.1.1 Synchronous Clock Case Types

The following clock case types are used in the titles of Chaps. 7, 8, 9, and 10.

Mantel Clock A mantel clock is a domestic clock intended for display on a mantelpiece or on a piece of furniture. It has a dial that can be read from across the room. They are mostly used in reception rooms. This is the commonest type of case used for domestic synchronous clocks.

Bedside Clock A bedside clock is a clock intended for display on a bedside table, but can also be used as a desk clock. A bedside clock has a dial that can be read from a short distance. Alarm clocks are time switches but can be used as bedside clocks. Some other time switches can be used as bedside clocks.

Wall Clock A wall clock is a clock that is intended for display high up on a wall. They are mostly used in kitchens. A wall clock has a dial that can be read from across the room. A time switch that controls central heating can be used as a wall clock which has a dial that can be read from a short distance.

Granddaughter Clock A granddaughter clock is a longcase clock that is less than 157 cm tall (eHow 2014). It has a dial that can be read from across the room.

They became popular in the 1930s. They are mostly used in reception rooms and in entrance halls. This is the least common type of case used for domestic synchronous clocks.

Novelty Clock A novelty synchronous clock is a clock that is primarily an ornament. In general, synchronous novelty clocks are not suitable for use as domestic clocks.

4.2 Synchronous Clock Case Styles

Existing books on domestic synchronous clocks do not include analysis of clock case artistic styles. Earlier, more general analyses are too general and too critical to be useful. Anonymous (1939) emphasised the importance of clock case design in stimulating sales and made the point that the design of the hands was particularly important. Anonymous (1943) also emphasised the importance of clock case design but stated that. 'Domestic clock-case design has drifted, to put it mildly, during the last generation'.

In 1936 a lecture on 'Modern Art in Clock Design' was given to the Horological Institute (Seager 1936a) The lecture was illustrated by examples, but no illustrations of clocks are included in the report of the lecture, and only one, an Art Deco clock, in the report of the subsequent discussion (Seager 1936b). There was no mention of Art Deco, or of other artistic styles. The importance of dial legibility is emphasised. There was much criticism of Art Deco features, especially the use of chrome for hands. However, the clock illustrated in the report is an Art Deco mantel clock with the caption. 'The height of perfection in modern clock design'. It is a Smith *St Edmund* synchronous mantel clock with a date range of 1935–1936 (Smith 2008).

The present analysis is based on the use of artistic and other styles as categories. Occasionally, a clock could equally well be placed in either of two categories. The analysis does not include all the clocks in Chaps. 7, 8, 9, and 10 since some clocks do not have a discernable style. Some domestic synchronous with discernable styles are illustrated in Sects. 4.2.1, 4.2.2, 4.2.3, 4.2.4, 4.2.5, 4.2.6, 4.2.7, 4.2.8, 4.2.9, 4.2.10, 4.2.11, and 4.2.12. In considering their artistic merits it must be remembered that all the synchronous clocks that have survived to be described in this book were initially bought by people who, presumably, liked them.

The analysis has two starting points. The first was the realisation that some domestic synchronous clocks have Art Deco cases. The second is the domestic clock case categories used by Smith (2008). His book includes numerous pictures of mechanical clocks, synchronous electric clocks, and other types of electric clock. Smith's categories, in the order used, are: Classic and Reproduction Clocks, Carriage Clocks, Longcase Clocks, Ships Clocks, Pictorial and Animated Clocks, Alarm Clocks, Travel Clocks, Timers, Plain Wall Clocks, English Clock Systems Clocks, Fancy Wall Clocks, Clocks with Rectangular Bezels, Wood Rectangular Case Clocks, Wood Early Style Clocks, Typical Wood Mantel Clocks of the

Twentieth Century, Wood Later Style Clocks, Rectangular Metal Case Clocks, Metal Non Rectangular Clocks, Bakelite Case Clocks, Plastic Case Clocks, Glass and Mirror Case Clocks, and Clock Cases of Other Materials. A large number of categories was needed to arrange the domestic clocks into groups with a common theme.

4.2.1 Art Deco Clocks

Art Déco is an artistic and design style that originated in Paris in the 1920s (Hillier 1968; Hillier and Escritt 1997; Pook 2014; Wikipedia 2011). The term first became popular in 1926, and it was soon anglicised to Art Deco. The style became widespread during the 1930s. It influenced all areas of design, including consumer items. Art Deco was regarded as representing a modern design style, with emphasis on the function of the object. It is based on symmetrical geometric shapes, using materials such as aluminium, stainless steel, lacquer, Bakelite, chrome, and inlaid wood. Stepped forms, geometric curves, and chevron patterns, are typical of Art Deco. By 1968 the Art Deco style had become debased. The application of Art Deco to clocks was logical since domestic clocks are often acquired as much for their appearance as for their timekeeping. In the UK, manufacture of domestic synchronous clocks also became widespread in the 1930s. At that time, synchronous clock movements were a new technology in the UK. It is therefore not surprising that, with its emphasis on function and modernity, that the Art Deco style was sometimes used for domestic synchronous clock cases. The Art Deco style spread widely through mass production (Hillier 1968) and its use in synchronous clocks is a typical example.

The four examples of domestic Art Deco clocks shown in Fig. 4.1 are of three different types. The flamboyant Temco synchronous mantel clock (Figs. 4.1a and 7.51) with its stepped form, cross banding, elegant Roman numerals and pierced hands is high Art Deco. It was made not later than 1938. The Smith *Delta* synchronous bedside clock, date 1933, (Figs. 4.1b and 8.16) is more restrained, but the geometric outline, flared sides, elegant Arabic numerals and geometric seconds indicator are all Art Deco. The Temco synchronous mantel clock shown in Figs. 4.1c and 7.48 is another flamboyant Art Deco clock, its stepped form is enhanced by a brass statuette on top and brass decorations on the front. The Ferranti Model No. 151 synchronous granddaughter clock (Figs. 4.1d and 10.1) is slim and elegant. The cross banded detailing on either side of the dial, elegant Roman numerals and chrome, pierced hands are all Art Deco.

4.2.2 Bakelite Clocks

Bakelite was a modern material in the 1930s. From a manufacturer's viewpoint it had the advantage that Bakelite clock cases were much cheaper to produce than

Fig. 4.1 Art Deco clocks, (**a**) Temco synchronous mantel clock, (**b**) Smith *Delta* synchronous bedside clock, (**c**) Temco synchronous mantel clock, (**d**) Ferranti Model No. 151 synchronous granddaughter clock

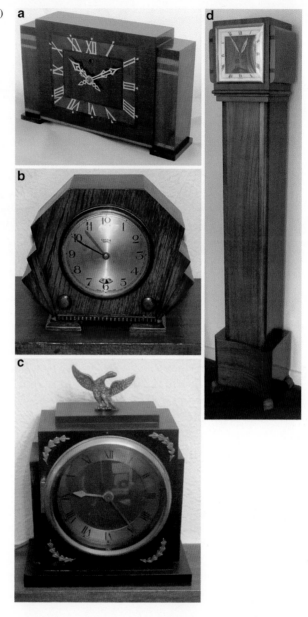

wooden clock cases (Lines 2012). The need to make tooling for moulding Bakelite restricted possible forms so Bakelite domestic synchronous clock cases tend to have a characteristic style that includes austere Art Deco. Two examples of Bakelite clocks are shown in Fig. 4.2. The simple curves on the top of the Synchronomains synchronous bedside clock shown in Figs. 4.2a and 8.32 are a characteristic feature

Fig. 4.2 Bakelite clocks, (**a**) Synchronomains synchronous mantel clock, (**b**) Temco synchronous mantel clock

of some Bakelite clocks. The clock was produced in the early 1930s. The Temco synchronous mantel clock (Figs. 4.2b and 7.53) is a Bakelite clock that is also pure Art Deco. It is an austere version of the clock shown in Fig. 4.1c. The clock was produced not later than 1938.

4.2.3 Carriage Clocks

A carriage clock is a traditional type of travel clock which has a balance wheel mechanical movement, a rectangular brass case, and a hinged handle on top. It is equipped with a leather travel case with a panel that can opened to view the dial without taking the clock out of its case. The example shown in Figs. 4.3 and 8.29 is a reproduction Carriage clock, without a travel case, that cannot be used for travel because of its mains lead. It is a Smith *Phillipe* synchronous bedside clock. The style of the dial is a usual feature of carriage clocks.

4.2.4 Chinoiserie Clocks

There was a vogue for Chinoiserie interior furnishing in the mid 1930s, and another in the mid 1950s. Chinoiserie domestic clocks, sometimes called lacquered clocks, were made in response. The nicely decorated Smith Chinoiserie synchronous mantel clock shown in Figs. 4.4 and 7.35 was made in 1936.

Fig. 4.3 Smith *Phillipe*
Carriage Clock Style
synchronous bedside clock

Fig. 4.4 Smith Chinoiserie
synchronous mantel clock,
(**a**) Front, (**b**) Top

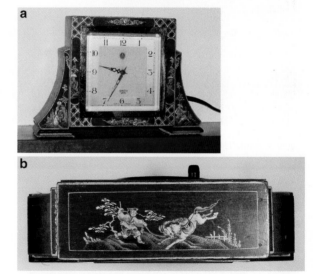

4.2.5 Commemorative Clocks

Synchronous commemorative clocks do not have a particular artistic style. Their
purpose was to commemorate various events. When only one clock was needed
this was done by the addition of a plaque. The Alexander Clark synchronous

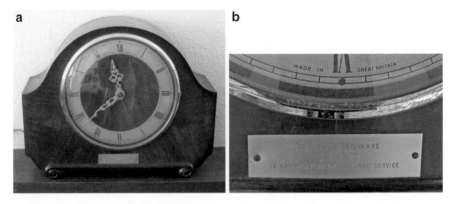

Fig. 4.5 Alexander Clark Commemorative synchronous mantel clock, (**a**) Front, (**b**) Presentation plaque

Fig. 4.6 Smith Egyptian Style synchronous mantel clock

mantel clock, shown in Figs. 4.5 and 7.1, is an example. The inscription on the commemorative plaque is 'BRITISH RAILWAYS C. Scott IN APPRECIATION OF 41 YEARS SERVICE'.

4.2.5.1   Egyptian Style Clocks

Specially commissioned synchronous clocks were produced when a number were needed. An example is the Smith Egyptian Style synchronous mantel clock shown in Figs. 4.6 and 7.36. The Egyptian style animals suggest that the clock was made to commemorate a Tutankhamen exhibition, probably the one held in 1952. It is an imposing clock, although the dial is a little too small for a mantel clock. It is also an example of the use of stone in synchronous clock cases.

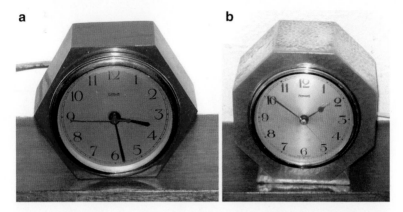

Fig. 4.7 Geometric clocks, (**a**) Goblin Model 394 Geometric Hexagonal synchronous bedside clock, (**b**) Ferranti Model No. 12 Geometric Octagonal synchronous bedside clock

4.2.6 Geometric Clocks

Some synchronous clock cases have a geometric outline, usual a regular hexagon or a regular octagon. These are debased Art Deco in that the geometric shape is made the raison d'etre rather than a feature of the clock case. Two examples are shown in Fig. 4.7. The Goblin Model 394 Geometric Hexagonal synchronous bedside clock (Figs. 4.7a and 8.10) has an onyx case. The Ferranti Model No. 12 Geometric Octagonal synchronous bedside clock (Figs. 4.7b and 8.5) has a hammered pewter case, with the base widened for stability. In neither case has the use of an exotic material made up for the gimmicky design.

4.2.7 G-Plan Style Clocks

G-Plan was a pioneering range of furniture in the United Kingdom, produced by E Gomme Ltd of High Wycombe (Wikipedia 2014). In 1953, Donald Gomme, the designer at E Gomme, decided to produce a range of modern furniture for the entire house. It was well made with unconventional, sometimes organic, shapes of smooth form. The name was coined by Doris Gundry of the J. Walter Thompson advertising agency and, in the 1950s, became regarded as definitive modern furniture. Some synchronous clocks were produced in a similar style. A Smith *Woburn Variant* G-Plan Style synchronous mantel clock is shown in Figs. 4.8 and 7.40. This is an example of a clock where the movement cover projects behind the rest of the case to give the illusion that it is a shallow case clock.

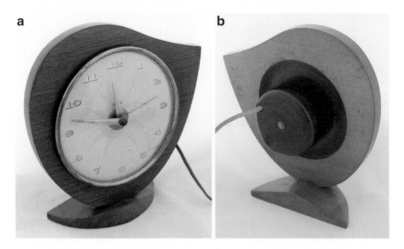

Fig. 4.8 Smith *Woburn Variant* G-Plan Style synchronous mantel clock, (**a**) Front, (**b**) Back

Fig. 4.9 Smith *Norwich*
Napoleon's Hat Style
synchronous bedside clock

4.2.8   Napoleon's Hat Clocks

Napoleon's Hat clocks are a traditional domestic clock case style with a wide base
and a smoothly rounded top. The height reduces towards the ends. Its traditional use
for mechanical clocks was extended to synchronous clocks. The example shown
in Figs. 4.9 and 8.17 is a Smith *Norwich* synchronous bedside clock, date range
1932–1933.

4.2.9   Pictorial Clocks

Synchronous pictorial clocks do not have a particular artistic style. Many pictorial
clocks are novelty clocks in which a small dial is placed somewhere in a picture.

Fig. 4.10 Smith *Huntingdon*
Pictorial synchronous mantel
clock

However, there are some in which a picture is nicely integrated with the dial design.
A Smith *Huntingdon* Pictorial synchronous mantel clock, date 1953, is shown in
Figs. 4.10 and 7.44. The easily read dial is integrated with a realistic hunting scene.
It is also a Bakelite style clock.

4.2.10   Utility Clocks

In 1943, during World War II, furniture was part of the rationing system in the
United Kingdom; and the Board of Trade set up the Utility scheme which limited
costs and the types of furniture on sale (Wikipedia 2014). A small number of simple
designs became available in oak or mahogany. The scheme ended in December
1952. It was suggested in 1943 that domestic clock cases needed to re-designed
to match this style of furniture (Anonymous 1943). In practice this did not happen
with domestic synchronous clocks. Some case designs from before the war were
continued, and most new designs were not noticeably utilitarian. However, some
wall clocks of simple Utility Style were produced. For example, an English Clock
Systems synchronous wall clock is shown in Figs. 4.11 and 9.1.

4.2.11   Wrought Iron Clocks

Wrought Iron Style clocks have cases that mimic the style of wrought iron fences
and gates, rather than furniture. They were first marketed in 1940 (Anonymous
1940). A Temco No. 1200 Wrought Iron Style synchronous mantel clock is shown
in Figs. 4.12 and 7.54. A similar Smith clock has a date range of 1956–1961
(Smith 2008).

Fig. 4.11 English Clock
Systems Utility Style
synchronous wall clock

Fig. 4.12 Temco No. 1200
wrought iron synchronous
mantel clock

4.2.12 Novelty Clocks

Synchronous Novelty Clocks are primarily ornaments and do not have a particular
artistic style. The Smith *Dickens* synchronous Novelty clock shown in Fig. 4.13
is an example. The date range is 1953–1955. It is primarily a picture but the
dial is not integrated into the picture. Further, the 5.5 cm diameter dial is too
small and too inconspicuous for the clock to be used as a mantel clock, and the
24 cm high × 33.5 cm wide case is too large for it to be used as a bedside clock.

Fig. 4.13 Smith *Dickens* synchronous Novelty clock

References

Anonymous (1935) Smith's clocks. Horological J 78(926):11

Anonymous (1936) Smith clocks. Horological J 78(934):9

Anonymous (1939) Current domestic clock design. Horological J 81(970):186–195

Anonymous (1940) Temco synchronous electric clocks. Horological J 82(984):14

Anonymous R (1943) Clock-case design. Opportunity. Horological J 85(1019):183

eHow (2014) http://www.ehow.co.uk/info_12269131_wages-1930s.html. Accessed 2014

Hillier B (1968) Art Deco of the 20s and 30s. Studio Vista, London

Hillier B, Escritt S (1997) Art Deco style. Phaidon, London

Lines MA (2012) Ferranti synchronous electric clocks, Paperback edition with corrections. Zazzo Media, Milton Keynes

Pook LP (2014) Temco Art Deco domestic synchronous clocks. Watch Clock Bull 56/1(407): 47–58

Seager JH (1936a) Modern art in clock design. Lecture delivered before the Horological Institute, on April 8th. Horological J 78(932):12, 14–17

Seager JH (1936b) Modern art in clock design. Discussion following lecture delivered before the Horological Institute, on April 8th. Horological J 78(933):12–15

Seager JH (1937) British clock cases. Horological J 80(949):70–72

Smith B (2008) Smiths domestic clocks, 2nd edn. Pierhead Publications Limited, Herne Bay

Wikipedia (2011) Art Deco. http://en.wikipedia.org/wiki/Art_Deco. Accessed 2011

Wikipedia (2014) http://en.wikipedia.org. Accessed 2014

Chapter 5
Servicing of Synchronous Clocks

Abstract Servicing of domestic synchronous clocks is not financially viable for professional clockmakers so servicing usually has to be carried out on a do it yourself (DIY) basis. This requires both clockmaking and electrical tools and skills. There are so many variables that it is only possible to give general guidance. New original spare parts are in general not available. Synchronous movements removed from clocks that have been scrapped or converted to quartz are sometimes available as a source of used original spare parts. Otherwise, the only source is to sacrifice a synchronous clock. Some items are available from horological suppliers. These include round and square clock glasses, hand set and starter knobs, and taper pins. BA threads are used in most British domestic synchronous clocks so replacement nuts and bolts can be obtained from horological or engineering suppliers. A systematic approach to fault finding is needed. Faults may be due to fair wear and tear or be the result of tampering. One clock examined had six distinct faults, all due to tampering. Provided that the motor coil is intact, cleaning and lubrication, and perhaps replacement of the mains lead, is usually all that is needed to get a synchronous clock that has stopped working back into running order. If there no apparent electrical faults are apparent, and a synchronous clock does not work, then it will have to be dismantled to find and rectify the fault or faults. Normal clock making techniques may be used.

5.1 Introduction

Current UK electrical regulations mean that servicing of domestic synchronous clocks is not financially viable for professional clockmakers so servicing usually has to be carried out on a do it yourself (DIY) basis. This requires both clockmaking and electrical tools and skills. Britten (1978) includes advice on specialist horological tools. These are available from horological suppliers. A good general servicing guide was published by Robinson (1940). In this chapter his advice has been updated where needed, expanded on the basis of experience, and on information in Philpott (1934a, b, 1935), Robinson (1942) and Stott (1946a, b). Much of the advice available on the Internet is misleading so this source should be used with caution.

© Springer International Publishing Switzerland 2015   71
L.P. Pook, *British Domestic Synchronous Clocks 1930–1980*, History
of Mechanism and Machine Science 29, DOI 10.1007/978-3-319-14388-0_5

New original spare parts for British domestic synchronous clocks are in general not available. Synchronous movements removed from clocks that have been scrapped or converted to quartz are sometimes available as a source of used original spare parts. Otherwise, the only source is to sacrifice a synchronous clock with a working movement. Some items are available from horological suppliers. These include round and square clock glasses, hand set and starter knobs, and taper pins. BA threads are used in most British domestic synchronous clocks so replacement nuts and bolts can be obtained from horological or engineering suppliers.

Provided that the motor coil is intact, cleaning and lubrication, and perhaps replacement of the mains lead, is usually all that is needed to get a synchronous clock that has stopped working back into running order. It is sometimes impossible to repair damage due to tampering by a previous owner (Pook 2014). Early synchronous clocks are usually well made so dismantling and re-assembly is straightforward. Some synchronous clocks were designed as cheap throwaway items, and servicing is difficult or impossible (Moore 1936).

A synchronous clock that does not run correctly is peculiarly irritating. To the lay observer there is little that can be done to put things right, apart from ensuring that electricity is reaching the clock. A systematic approach to fault finding is needed. Faults may be due to fair wear and tear or be the result of tampering. One clock examined had six distinct faults, all due to tampering.

5.2 Electrical Faults

Previously published advice on servicing synchronous electric clocks is based on the implicit assumption that a synchronous clock that was previously in running order has stopped working correctly. However, an old synchronous clock that has been acquired is often in unknown condition. It may have been tampered with by a previous owner. It might have been sold as for spares or repair, with the lead either cut off, or removed completely. The UK mains voltage is potentially lethal so a cautious approach is needed. No old synchronous electric clock should simply be plugged in to see whether it works, even if it has been stated by a seller to be in working order.

The first check needed is to use a meter to check the continuity and DC resistance of the coil. This should be around 2,000–3,000 Ω. A low DC resistance usually means that there is a short circuit within the coil. The clock might work but it would overheat. Unless there is a fault within the wiring leading to the coil an open circuit means that there is break within the coil or that one end of the coil wire has become detached. An open circuit in the coil is the second most common cause of a synchronous clock not working. The wire used in the coil is very fine, typically 0.07 mm diameter, and special purpose equipment is needed to repair the break or rewind the coil. The only practical DIY cure is to replace the coil. Rewinding the coil using thicker wire and running the clock at low voltage through a transformer is sometimes recommended. This works, but means that the clock is no longer original and so is of less interest to a collector.

If a mains lead is in poor condition or missing it should be replaced. In some clocks the movement cover has to be removed for access to the terminals. The movement cover is usually held in place by a screw at the back. Twin core 0.5 mm^2 cable, sold for use with portable lights, is suitable. It is small enough to fit through the holes provided in synchronous movement covers and synchronous clock cases. There is no point in replacing an old lead that is in good condition simply on the grounds that it does not meet current electrical regulations. This does not mean that the clock is unsafe to use. Fit a plug, if needed, and check the terminals for tightness. Some plugs are fused. The fuse should be of the correct rating, usually 3 A.

When the mains lead and plug are known to be in order, and the DC resistance is correct, then the clock may be plugged in to see whether it works. First check that the clock is for the available mains voltage. The permissible voltage range is nearly always marked on the back of the clock. Plug the clock in and turn the power on cautiously, being ready to turn off immediately if anything untoward happens. This is very rare, but if it happens switch off immediately and dismantle the clock to find and rectify the source of the trouble. Next, check any exposed metal parts, using an insulated electrician's screwdriver fitted with an indicator neon, to ensure that they are not live. It is very rare for exposed metal parts to be live.

If all is in order electrically a self starting clock might start automatically, or it might be possible to start a manual start clock. If the clock starts then watch it running for a few minutes. If all seems well leave it for at least 12 h to ensure that it continues to run, and keeps good time. If all seems well then the synchronous clock can be regarded as being in working order.

5.3 Mechanical Faults

If a synchronous clock does not work then it will have to be dismantled to find and rectify the fault or faults. Normal clock making techniques (Britten 1978; Rawlings 1993; Robinson 1940, 1942) may be used. There are so many variables that it is only possible to give general guidance.

5.3.1 Dismantling

Any synchronous clock will have been assembled in a particular order. Service manuals are not usually available, so the problem is to work out what the order of assembly was and reverse it. The progress of dismantling should be recorded in notes and by taking digital photographs. These are an essential guide to re-assembly. Many of the photographs of movements in Chap. 11 were taken during dismantling. Parts should be stored in labelled containers. A particular point to note during dismantling is that hand set knobs sometimes have left hand threads. Deterioration means that dismantling an old synchronous clock can be difficult, and

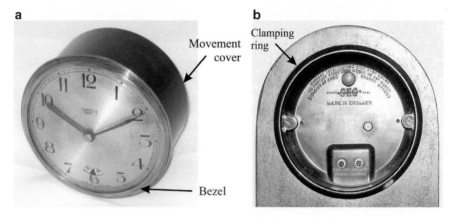

Fig. 5.1 Smith back set Type 1 movement, (**a**) as an insertion movement, (**b**) held in place by a clamping ring

patience may be needed to ease items apart. Nuts and bolts are sometimes very tight so well fitting tools are needed. Slack nuts and bolts, missing nuts and bolts, and bolts with damaged screwdriver slots are usually an indication of tampering.

It is usually easy to see how to remove a synchronous movement from the case, although doing so is sometimes fiddly. Some synchronous clocks are fitted with insertion movements. A Smith back set Type 1 movement (Sect. 11.13.1) as an insertion movement is shown in Fig. 5.1a. Back set refers to the position of the hand set knob. The glass and dial are held in place by a bezel. This has lugs on the back which engage with lugs on the front of the Bakelite movement cover. The bezel can be removed by twisting it anticlockwise. The insertion movement fits in a cylindrical hole in a solid wood clock case, and is held in place by a clamping ring (Fig. 5.1b).

There are two common dismantling problems with insertion movements. The first is that the wooden case has shrunk across the grain, jamming the movement cover. The insertion movement can usually be eased out forwards slowly by using a rocking movement at the back. Once out, enlarging the hole slightly so that the insertion movement is a sliding fit is straightforward woodwork. The second common problem is that the bezel is jammed. It can be freed by carefully bending the lugs backwards until the bezel is just free to rotate. Both these cures need patience. The movement is held in the movement cover by screws which can be removed when the dial has been removed.

It is usually necessary to remove the hands and dial to dismantle the movement. The hands are usually a push fit, and are sometimes very tight, so special tools are needed. Figure 5.2a shows a pair of home made hand levers being used to remove the minute hand from the Temco clock shown in Fig. 7.49. Hardboard offcuts are used as packing and to protect the dial. Figure 5.2b shows a hand removing tool being used to remove the hour hand. The small jars visible through a translucent lid are used to store clock parts. Older clocks sometimes have the hands held in place by a nut or a taper pin. Removing these to free the hands is straightforward.

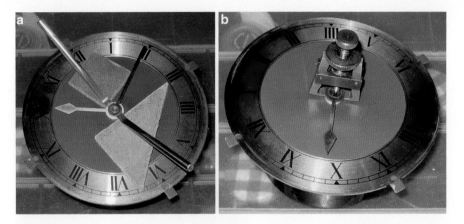

Fig. 5.2 Removing hands from a Temco synchronous mantel clock, (**a**) hand levers being used to remove the minute hand, (**b**) hand removing tool being used to remove the hour hand

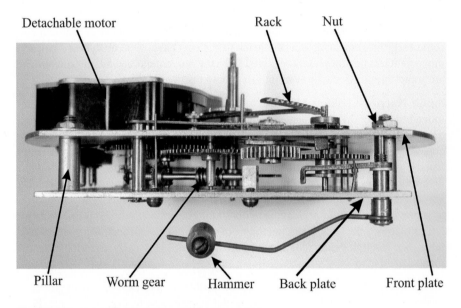

Fig. 5.3 Top view of Smith Narrow striking movement

When the movement is out of the case it can be examined for faults and to see how to dismantle it. The commonest mechanical fault that prevents the movement running is lubricant congealed or dried up, followed by the starting mechanism not operating correctly, incorrect assembly, and tampering damage. The detail design of synchronous movements varies widely, and two examples are shown in Figs. 5.3 and 5.4. Early movements are usually easy to dismantle. Some movements

Fig. 5.4 Rear view of Smith
QGEM movement

Motor   Twisted tab

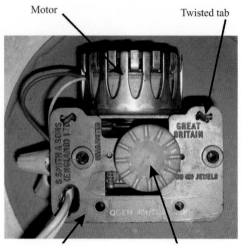

Plastic plate   Hand set knob

are built into movement covers, and have to be dismantled in layers. A problem with some early movements is that the leads to the coil have to be unsoldered before a movement can be completely removed from the movement cover. Later movements are sometimes difficult, or even impossible, to dismantle.

Smith Narrow Striking Movement Some features of a Smith Narrow striking movement (Sect. 11.13.10) are shown in Fig. 5.3. The brass plates are the two main members of the movement which support and locate the other parts. The plates are connected by four pillars, which are riveted to the back plate and fastened to the front plate by nuts. The detachable motor can be removed and replaced without disturbing other parts. The rack and hammer are part of the striking work. The reduction gear includes two worm gears. The front plate has to removed in order to remove wheels etc. for cleaning. This should be done carefully so that parts are left in place on the back plate and their position can be recorded. Worm gears have their own mountings, attached to one of the plates. This early movement is well made and easy to dismantle.

Smith QGEM Movement The sort of problem that can be encountered in dismantling some synchronous movements is illustrated in Fig. 5.4 which shows a rear view of a Smith QGEM movement (Sect. 11.13.8). This has been designed as a cheap throw away item. The motor is held in position by twisted metal tabs. It is possible to untwist these, but it is difficult to ensure that they are correctly re-twisted when re-assembly is by hand. It can done. Two faulty QGEM movements have been cannibalised to make one good movement.

5.3.2 *Cleaning and Oiling*

Published advice on cleaning synchronous movements is conflicting. From a DIY viewpoint dry methods based on clean soft rag, pegwood and elbow grease are the best. Successive pieces of clean rag can be used on flat and curved parts until the rag comes away clean. Pegwood is small sticks of close grained wood, available from horological suppliers. It is used principally for cleaning out pivot holes. It is sharpened to a point to enter the pivot hole and twirled round to remove the dirt. This is repeated until the pegwood comes out clean. Pegwood can also be used to clean dirt from gear teeth. Cleaning fluids and cleaning machines should be regarded as a last resort. Fluids do not mix well with electricity, and can damage some finishes. If they are used then makers' instructions must be followed.

It is not always possible to dismantle moving parts completely. For example, the rotor arbor of Temco Mark IV and Mark V movements (Sects. 11.15.3 and 11.15.4) cannot be removed from its bearings because the starter knob is force fitted onto one end, and a pinion force fitted on the other end. One method of cleaning is to flood the bearings with clock oil and wipe away the surplus. This is repeated until the surplus oil is clean. This method might appear to be poor practice but, at the time of writing, a Temco clock with a Mark IV movement, on which this was done, has been running for 5 years without further attention to the movement.

Published advice on the lubrication of synchronous movements is conflicting. Two manufacturer's recommendations are known. The first is that oil was supplied by Energol for Smith synchronous clocks with sintered bearings (Fig. 5.5). This looks and smells like a mineral motor car engine oil. The second is that a very

Fig. 5.5 Energol oil for Smith clocks

light motor oil was recommended by Vernor (1945) for Westclox synchronous clocks. Experience shows that a modern synthetic clock oil is satisfactory. This should be applied sparingly using a clock oiler, available from horological suppliers.

5.3.3   Re-assembly and Testing

Re-assembly of synchronous movements can be fiddly, but does not usually present serious difficulty. The main point is to ensure that all moving parts can move freely. All gear pairs must have the correct depth (of engagement). Wheels pivoted in plates will automatically have the correct depth, provided that wear is not excessive. Some worm gears have provision for adjusting the depth, and this should be checked. Detachable motors do not normally have means of positive location, so the gear depth should be checked. There is usually enough clearance in fastener holes for the depth to be adjusted before the fasteners are tightened. A binding output pinion is a common fault in synchronous movements that have been tampered with. Similarly, in a non detachable motor, adjustment may be needed to ensure that the rotor clears the stator. Check that the friction drive to the hands feels firm. If necessary tighten the spring. Occasionally, oil makes the friction drive slip. This sometimes happens when the movement has been flooded with oil in a misguided attempt to make it run. The oil will have to be removed.

Replacing a plate and engaging pivots in the plate holes needs a systematic approach. Referring to Fig. 5.3, the back plate should be firmly supported, but with clearance for parts that project below the plate. Assemble the wheels etc. in the pivot holes in the back plate. A photograph taken during dismantling to show their correct location is useful. Gently lower the front plate onto the pillars as far as it will go. If possible, start nuts that hold the front plate in position. With gentle pressure on the plates manoeuvre the pivots, one at a time, into their holes. The order in which this needs to be done is usually obvious. Special tools are available, but pegwood or a small screwdriver can be used. When all the pivots appear to be engaged, lightly tighten the nuts and check that all the pivoted arbors have end float. If not, correct as needed and fully tighten the fixing nuts. Complete re-assembly and oiling, except for fitting the dial and hands. Ensure that all moving parts are free to move. Check that the rotor rotates freely when a manual start is operated. Then bench test the synchronous movement before casing up.

Bench testing must be done with caution because there will usually be exposed live parts. Connect the power. A self starting synchronous movement should start automatically, and it should be possible to start a manual start movement. If the movement does not start use a meter to ensure that the voltage across the coil is correct. If this is in order watch the starter mechanism in action to see exactly what is happening: adjustment of parts is sometimes needed. If the movement runs leave it running for at least 12 h. This means that the gears will have gone through all possible positions.

Stopping at regular intervals after the movement has been started means that something, usually dirt or a damaged tooth, is catching somewhere. Finding out where, and rectifying the fault, may need patience. If a tooth is damaged replace the wheel. Stopping at irregular intervals usually means that there is excessive wear in the motor bearings. Ironically, this sometimes becomes apparent after dirt has been cleaned out of the bearings, and parts have to be replaced.

When all is in order, fit the dial and hands, and run for 12 h to make sure that the hands clear each other, then case up. Fitting the hands may have to be left until after casing up. Some dials have a small notch at 6 o'clock to ensure that the dial is correctly located. Make sure that this engages with the corresponding projection. The only problem likely to be encountered is a loose push fit hand. Some hour hands have a split collet. The hour hand from a Smith granddaughter clock (Fig. 10.3) is shown in Fig. 5.6. If necessary the split collet can easily be tightened. A plain collet can usually be tightened by passing a single strand of fine wire through the hole in the hand before pushing it into position. Some minute hands have a square hole which fits on a square on the arbor. There are four ways of fitting the minute hand onto its arbor. For a striking or chiming clock only one of these is correct.

With the movement running, and the hands in position, watch and listen to striking and chiming work to ensure that it is operating correctly. If it is out of sequence, adjustment is needed. It is usually easy to see what needs to be done.

Once running, a synchronous clock usually keeps good time, but will sometimes loose a minute or so per day. This can be due to a slipping friction drive (Philpott 1934b) but this should not happen in a carefully serviced clock. The other reason is that the motor is running at just below the synchronous speed (Sect. 3.2.3.2). The probable cause is excessive friction somewhere, but this is difficult to trace.

Fig. 5.6 Split collet on the hour hand of a Smith granddaughter clock

Split collet

References

Britten FJ (1978) The watch & clock makers' handbook, dictionary and guide. 16th edn. Revised by Good R. Arco Publishing Company, New York

Moore GE (1936) Miniature synchronous motors in service. Horological J 78(933):3–4

Philpott SF (1934a) Synchronous clocks – troubles and remedies. Horological J 77(911):88, 90

Philpott SF (1934b) Synchronous clocks – troubles and remedies. Horological J 77(914):55–56

Philpott SF (1935) Modern electric clocks, 2nd edn. Sir Isaac Pitman & Sons Ltd, London

Pook LP (2014) Temco Art Deco domestic synchronous clocks. Watch Clock Bull 56/1(407): 47–58

Rawlings AL (1993) The science of clocks and watches, 3rd edn. British Horological Institute Ltd., Upton

Robinson TR (1940) Servicing the synchronous clock. Horological J 82(984):13–36

Robinson TR (1942) Modern clocks. Their repair and maintenance, 2nd edn. N A G Press Ltd., London

Stott HT (1946a) Electricity and horology. Horological J 88(1048):23–30

Stott HT (1946b) Synchronous motors. Horological J 88(1050):110–111

Vernor JH (1945) Oiling synchronous clocks. Horological J 87(1039):116

Chapter 6
Analysis of Marketing and Reliability

Abstract Synchronous electric clock technology consists of two related technologies. One is small synchronous electric motors. The other is a reliable AC mains supply of accurately controlled frequency. By 1930 both of these were mature technologies in America. AC mains supplies of accurately controlled frequency started to become available in the UK in the early 1930s, and there was immediate interest by British manufacturers in the production of domestic synchronous clocks. There are five identifiable phases in the sales of British domestic synchronous clocks. In the UK, domestic synchronous clocks were a new must have technology, and sales boomed during the 1930s. Government restrictions on production led to a lull in sales during the Second World War (1939–1945). In the late 1940s pent up demand for domestic clocks of all types following war time restrictions led to a second sales boom. Sales were broadly level from about 1950 to the late 1960s. Domestic synchronous clock sales declined from the late 1960s onwards and never recovered. By 1980 it was all over. Analysis of possible reasons for the final decline in sales shows that this was due to the poor overall reliability of domestic synchronous clocks. In the 1970s pent up demand for domestic clocks had been satisfied so there was no longer an external driver for sales that overcame the poor overall reliability. Other factors were of secondary importance.

6.1 Introduction

Synchronous electric clock technology actually consists of two related technologies. The first technology is small synchronous electric motors suitable for incorporation in synchronous clock movements. The second technology is a reliable AC mains supply of accurately controlled frequency. By 1930 both of these were mature technologies in America. AC mains supplies of accurately controlled frequency started to become available in the UK in the early 1930s and there was immediate interest by two British manufacturers in the production of domestic synchronous clocks. One, Smiths Industries, was a manufacturer of domestic mechanical clocks (Smith 2008). The other, Ferranti Ltd was an electrical company (Lines 2012). Other entrants soon followed, and production of synchronous clocks increased up to the outbreak of the Second World War in 1939. Production decreased substantially during the war. It resumed on a large scale after the end of the war in 1945, and

© Springer International Publishing Switzerland 2015

L.P. Pook, *British Domestic Synchronous Clocks 1930–1980*, History of Mechanism and Machine Science 29, DOI 10.1007/978-3-319-14388-0_6

there were several new entrants. Only one of these, Metamec, became a major manufacturer of synchronous clocks (Bird 2003). The British subsidiary, Westclox Ltd, of an American company also became a major manufacturer (Linz 2004).

There were three main stakeholders in British domestic synchronous clocks.

1. Manufacturers of British domestic synchronous clocks who were interested in the following.

    (a) Marketing, including brands.
    (b) Technical aspects of movements, including patents and production costs.
    (c) Clock case design, including production costs.
    (d) Distribution to the user.
    (e) The legal environment.

2. Users of British domestic synchronous clocks who were interested in the following.

    (a) Timekeeping.
    (b) Reliability.
    (c) Appearance.
    (d) Availability.
    (e) Cost.

3. Regulatory authorities who were interested in the following.

    (a) Trading standards.
    (b) Safety.
    (c) Control of the AC mains supply.

Electricity supply companies did not regard themselves as stakeholders because the value of electricity used by synchronous clocks is insignificant.

There are five identifiable phases in the sales of British domestic synchronous clocks (Sect. 2.1.2).

1. In the UK, domestic synchronous clocks were a new must have technology, and sales boomed during the 1930s.
2. Government restrictions on production led to a lull in domestic synchronous clock sales during the Second World War (1939–1945).
3. In the late 1940s pent up demand for domestic clocks of all types following war time restrictions led to a second sales boom for domestic synchronous clocks. New manufacturers appeared to satisfy the demand. Manufacturers saw a different driver for a boom. There was a belief that they were still a must have technology. For example, Barrett (1946) noted. 'That these clocks have a very bright future there is no doubt, in fact they are the clocks of the present and future as far as can visualised especially when one considers that a further £150 million is to be spent on the National Grid. . . . In fact these clocks will revolutionise the horological field.'
4. Sales of domestic synchronous clocks appear to have been broadly level from about 1950 to the late 1960s.

5. Domestic synchronous clock sales declined from the late 1960s onwards and never recovered. By 1980 it was all over.

Manufacturers always regarded declining sales as due to external factors. For example, when sales declined in 1967/1968 Westclox Ltd believed this was primarily due to competition from artificially cheap clocks imported from the then Iron Curtain countries (Wikipedia 2014). In an attempt to maintain the status quo, strong petitions to the UK Government produced the passing of an anti-dumping law and sales picked up, but this did not last. Another view was that the unanticipated arrival of quartz clocks as a cheap new technology was the reason for declining sales (Smith 2008).

Domestic mechanical clocks are a centuries old technology. Sales of British domestic synchronous clocks displaced some sales of mechanical clocks but did not replace them. Sales of both types continued in significant quantities. For example, see Bird (2003). Sales of both types declined with the introduction of domestic quartz clocks. Domestic mechanical clocks have been largely replaced by domestic quartz clocks, but sales of domestic mechanical clocks have continued as an upmarket niche product. Domestic synchronous clocks were different. They have been completely replaced by domestic quartz clocks.

In this chapter analysis of the marketing and reliability of British domestic synchronous clocks leads to an explanation of the rise and fall of the technology.

6.2 Analysis of Marketing

For consumer goods, such as domestic clocks, marketing has two different meanings (Wikipedia 2014). The context will usually show which is intended.

1. Marketing can be regarded as an organisational function and a set of processes for creating and distributing goods to consumers, and improving the user experience. Marketing is also the science of choosing target markets through market analysis, as well as understanding consumer behaviour, and providing superior customer value.
2. Marketing is the process of distributing goods to consumers.

6.2.1 Marketing Strategies

In retrospect, the marketing of British domestic synchronous clocks, in the first sense, was mostly based on five strategies.

1. Establishing strong brands so that prospective purchases were led to believe that they would have a good user experience with timekeeping and reliability. For example, Anonymous (1940a) noted that.

The policy of most synchronous clock manufacturers was to brand the clocks and sell them at maintained prices. It is probable that without the invention of the synchronous clock, the UK today might have been without a clock-making industry. The compactness of the movement and its adaptability to fitting in cases of every type and in novel positions score tremendous advantages over the mechanical clock. Clocks can be mounted in comparatively inaccessible positions, lending themselves to incorporation in novel schemes of decoration.

Some contemporary advertisements for domestic synchronous clocks emphasised their reliability. Further, the offer of a guarantee, usually for 1 year, was an implicit claim that a domestic synchronous clock was reliable. This level of reliability was not difficult to achieve because, by the time they were introduced in the UK, synchronous clocks were already a mature technology.

2. Producing a wide range of clock cases so that more prospective purchases would be able to locate clocks that were to their taste (Sect. 2.1.2).

3. Maintaining prices at artificially high levels to protect the profit margins of manufacturers, wholesalers, and retailers, who were usually jewellers (Sect. 2.1.1).

4. Restricting competition from imported clocks. For example, Anonymous (1940b) noted that.

> Members of the British Synchronous Clock Conference have, however, received wonderful co-operation from distributors – handling only electric clocks made by members of the Conference and not allowing foreign electric clocks to get into general distribution in this country.

5. Distribution (marketing in the second sense) was usually made the responsibility of a separate company, sometimes a subsidiary of the manufacturer.

6.2.2   Technical Aspects

It is technical aspects of the synchronous motors used in the movements of British domestic synchronous clocks that are of concern to manufacturers. The motors are of two distinct types: reluctance motors, which have a non magnetised rotor, and magnetised motors, which have a magnetised rotor. Both types were made in manual start and self starting versions (Sect. 3.2). There were no significant changes in the technology of synchronous motors produced by British manufacturers during the period covered by this book (Sect. 11.17). There were, of course, detail changes to improve reliability and reduce production costs, including the use of plastics. Many of the British Patents relevant to synchronous clocks include 'Improvements' in their title.

The designer of a synchronous movement has to avoid infringing patents. Some apparently eccentric features appear to have been incorporated to avoid infringing a patent. Avoiding infringement was a problem for early British manufacturers since some patents were American owned. In particular, the shaded pole method of self starting (Sect. 3.2.4.4) could not be used. This method is superior to other self starting methods because there are no moving parts. There do not appear to have been any major disputes over patents. Occasionally, patents owned by another manufacturer are acknowledged on clocks.

Magnetised motors only appear to have been produced by major British domestic synchronous clock manufacturers, presumably because high development costs could be amortised over a large number of units (Sect. 11.17). Minor manufacturers only produced reluctance motors, and in general their movements do not appear to use features that were protected by patents.

6.2.3 Clock Case Design

Clocks for domestic use have been produced for four centuries. Historically, clock case design styles have been based on furniture design styles (Sect. 4.1) with the obvious constraints that the case must be big enough to enclose the movement, the dial must be legible, it must be possible to set the hands, and only available materials, usually wood, could be used. Cases were usually no much bigger than needed to enclose the movement. Addition requirements for synchronous movements are provision for the mains lead and, for non self starting movements, access to a starting knob or lever. With the exception of Art Deco clocks, designers of British domestic synchronous clock cases were unadventurous.

A very large number of clock case designs were used for British domestic synchronous clocks. Those illustrated in Chaps. 7, 8, 9, and 10 are only a small selection which cannot be regarded as a random sample. Many of them do not have any discernable style. Nevertheless, there are some discernable trends. In the 1930s most cases were wooden, and often based on traditional clock styles. In the late 1930s there was a trend to the use of shallow cases to suit narrow mantel pieces (Sect. 4.1). Although not a style it had a clear influence on synchronous clock case design. Also in the late 1930s there was a trend to the use of a wider range of materials such as stone, glass, mirror glass, and plastics. Synchronous clocks using these materials were sometimes described as *modern* in advertisements even though there was no discernable consistent style. In mechanical clocks visible parts of movements, usually pendulums and occasionally escapements, are sometimes used in clock case design. By contrast, visible movements never featured in domestic synchronous clock case design.

The only extensively used new clock case style was Art Deco (Sects. 4.2.1, 7.29, 8.21, and 9.16), and an extensively used new material was Bakelite (Sect. 4.2.2). Both these were new at much the same time as synchronous domestic clocks were a new technology in the UK. There is a remarkable parallel between the rise and fall of British domestic synchronous clocks, and the rise and fall of Art Deco (Pook 2014). With the resumption of large scale production after the Second World War British domestic synchronous clock case design followed on from where it left off, with some case designs carried over from before the war. Synchronous clock case design lost its way and became increasingly gimmicky (Sect. 4.2). The only discernable elegant new style was G-Plan (Sect. 4.2.7).

Some well designed British domestic synchronous clocks were produced. Three examples are shown in Fig. 6.1. Production figures for individual domestic synchronous clocks are not available, but a rough indication of sales can be obtained from the frequency with which surviving clocks are offered for sale on eBay. By this test the clocks shown in the figure were all, deservedly, best sellers. The Ferranti Model No. 3 synchronous bedside clock, shown in Fig. 6.1a, is a high Art Deco Bakelite clock, date range 1932–1938 (Sect. 8.4). The Smith *Nell Gwynne* synchronous bedside clock, shown in Fig. 6.1b, is a reproduction of a traditional

Fig. 6.1 Best selling clocks (**a**) Ferranti Model No. 3 synchronous bedside clock, (**b**) Smith *Nell Gwynne* synchronous bedside clock, (**c**) Temco synchronous bedside clock

style, date range 1950–1978 (Sect. 8.14). The clock is unusual in that, with the door open, the rotor is visible through the transparent movement cover. The Temco synchronous bedside clock, shown Fig. 6.1c, is restrained Art Deco, made not earlier than 1938 (Sect. 8.20).

6.2.4 Practical Aspects of Design

Practical aspects of design are those features that need to be well designed to ensure good user experience. For domestic synchronous clocks these include the following. A dial that can be easily read over a wide range of viewing angles. Provision of some sort of tell tale, perhaps a seconds hand, to make it easy to tell whether or not the clock is running. A significant proportion of British domestic synchronous clocks do not have a tell tale (Sects. 7.29, 8.21, 9.16, and 10.6). Easily accessible hand set and starter knobs. For wall clocks, a well designed hanger so that it is easy to install the clock. This might seem obvious, but one or more of these aspects were often overlooked by British domestic synchronous clock designers. This is a common situation in the design of everyday things such as clocks (Norman 2013).

6.2.5 Distribution

Distribution (marketing in the second sense) was usually made the responsibility of a separate marketing company. In retrospect this separation of marketing functions was not a good idea.

Marketing companies sold British domestic synchronous clocks to retailers either directly, or indirectly through wholesalers. In either case retailers selected a limited number of types that they believed would sell from the very large numbers that were available. This meant that, in practice, a customer visiting a retailer would find that only a small selection of synchronous clocks was available, with perhaps none that were to their taste. This did not matter to the retailer in boom times when customers were perhaps prepared to compromise in order to secure a synchronous clock.

6.2.6 Legal Framework

Due to political pressure, the legal framework within which British domestic synchronous clock manufactures and distributors operated became increasingly rigorous during the period within which they were produced. The changes restricted the ability to maintain prices and limit foreign imports, and hence maintain profit margins at artificially high level.

6.2.6.1 Retail Prices

The prices at which consumer goods, including domestic synchronous clocks, are sold in the UK depend primarily on the cost of production and distribution, but prices are also influenced by legislation (Wikipedia 2014). In the 1930s Minimum Retail Price Maintenance, usually known as RPM, was in force. Manufacturers could, and did, enforce the retail price at which British domestic synchronous clocks were sold. Prices in advertisements and manufacturers' catalogues were what you had to pay. They were expensive. Some prices are shown in Lines (2012). Typically, the price of a British domestic synchronous clock was a week's wages. Nevertheless, Barrett (1939) stated that. 'Notwithstanding the fact that it has not been pushed as it should by the sellers of clocks, no less than thirty-three and one third percent of all the money expended on clocks in this country is represented by the sale of the synchronous type.'

This situation did not last. The main purpose of the Resale Prices Act 1964 was to prohibit RPM except where it could be shown that it was in the interest of the public as consumers or users of the good in question (Wikipedia 2014). The Resale Prices Act 1964 was repealed in 2000 because its provisions were incorporated in the Competition Act 1998. RPM was progressively abandoned for various goods. The last to go was the Net Book Agreement which set retail prices for books. This was abandoned in 1997.

6.2.6.2 Foreign Imports

Historically, the UK government protected British manufacturers from foreign imports by the imposition of tarifs on imported goods and by the imposition of anti-dumping duty (ADD). Dumping is the sale of goods at a price considerably less than their *normal* value. This is defined as a price that is lower than the price of similar goods in the country from which they originate. Tarifs and ADD were sometimes imposed in response to petitions from British manufacturers.

The European Union (EU) is a politico-economic union of 28 member states, including the UK, that are primarily located in Europe (Europa 2014). The Maastricht Treaty established the European Union under its current name in 1993. The promotion of free trade was a founding principle of the EU in the belief that it stimulated economic growth. To this end the EU has developed a single market through a standardised system of laws that apply in all member states. Tarifs within the EU were completely eliminated by 1968. This meant that tarifs could not be applied to goods imported to the UK from other EU countries. This has had a profound effect on British manufacturers, but agile, innovative companies have continued to prosper.

A history of modern anti-dumping law starts with the 1947 GATT (General Agreement on Tarifs and Trade) (Davis 2009). Anti-dumping law in all jurisdictions comes under Article VI of the GATT, but how states investigate alleged dumping and impose ADD is open to wide discretion. This led to the establishment of the World

Trade Organisation (WTO) in 1994. The Anti-dumping Agreement 1994 is one of three World Trade Organization principal trade defence agreements (Wikipedia 2014). The EU, including the UK are a party to the agreement, which made it more difficult for the UK government to impose ADD. However this change was too late to have any effect on British domestic synchronous clock manufacturers.

6.2.7 Conclusions on Marketing

There is nothing in the analysis of the marketing of British domestic synchronous that suggests reasons for their fall. Similar criticisms apply to the marketing of British domestic mechanical clocks and British domestic quartz clocks. It is therefore necessary to look elsewhere for the reason for the fall.

6.3 Analysis of Reliability

The overall reliability of a synchronous clock may be measured as the mean time between failures. It is a combination of the reliability of the movement, and the reliability of the AC mains supply. To a user it is the overall reliability that is important. By contrast, the marketing strategy of British manufactures (Sect. 6.2.1) concentrated on the reliability of the movement. The failure to recognise the importance of the distinction was a marketing failure.

6.3.1 AC Mains Supply

From a synchronous clock user's viewpoint the AC mains supply to premises in which the clock is installed has three important features, none of which are under the control of either the user or the manufacturer. Firstly, the voltage must be correct within a fairly wide range. Typically clocks are marked 200–250 V. This is not a problem in the UK because AC mains supply voltages are within this range. Secondly, the frequency must be tightly controlled so that a clock keeps good time. Again, this is not a problem in the UK. Thirdly, the mains supply must be reliable. It has never been completely reliable. Statistics published in 1937 showed that interruptions to the mains electricity supply averaged one interruption per customer per year (Read 2012). Reliability decreased during the Second World and in the post war period. Falling sales were sometimes ascribed to poor reliability, and power cuts were cited as the reason for the failure of at least one minor British synchronous clock manufacturer. Recent experience shows that, typically, there are now two or three interruptions per year. In addition, there are three or four local interruptions due to circuit breakers blowing. Hence, there are now about six interruptions per year.

It is possible to fit a synchronous clock with a carry over mechanism which keeps it running during a power failure (Anonymous 1935). However, this increases production costs, and was never used in British domestic synchronous clocks.

6.3.2   Determination of Reliability

Formally, the reliability of a clock may be measured as the mean time between failures. It is a combination of the reliability of the power source, the timekeeping element, and the rest of the movement. A distinctive feature of a domestic synchronous clock is that the timekeeping element and power supply are from the same source. By contrast the timekeeping element of a domestic mechanical clock are distinct. The timekeeping element of a mechanical clock is an oscillator, sometimes a pendulum, which oscillates at constant frequency. The power source is usually either a falling weight, or a mainspring, which is periodically rewound. Reliability data in Sects. 6.3.2.1, 6.3.2.2, and 6.3.2.3 are based on personal experience with a number of clocks over many years. They are sufficiently accurate for useful conclusions to be drawn.

6.3.2.1   Power Source Reliability

The power source of a mechanical clock is a falling weight or a mainspring, and is under the control of a user, who can be expected to ensure that the clock is wound up regularly. Power source failures, such as a broken mainspring, are very rare, so the power source is very reliable, with a mean time between failures of at least 20 years. The power source in a domestic quartz clock is a battery which is replaced by the user as needed. Typically, a battery lasts about 3 years so the mean time between failures of the power source of a quartz clock is less than that of a mechanical clock, but better than that of a synchronous electric clock, which is about 2 months (Sect. 6.3.1).

6.3.2.2   Movement Reliability (Repair or Replacement)

The reliability of a domestic clock can also be defined as the mean time between failures of the movement. When a movement fails it either has to be repaired or replaced. Available data suggest that, using this definition, the mean time between failures of a good quality mechanical clock, a good quality synchronous clock, and a quartz clock are all about 10 years.

6.3.2.3   Movement Reliability (Replacement)

An alternative definition would be to define failure times as how long a domestic clock can be kept in use and in working order before major components, or the complete movement, have to be replaced. Available data suggest that, using this definition, the mean time between failures of a good quality mechanical clock is measured in decades or sometimes centuries, and that of a good quality synchronous clock is measured in decades. However, the mean time between failures of a quartz clock is about 10 years. Alarm clocks are less reliable because of the added complexity, and the mean time between failures is about 5 years.

6.3.2.4   Overall Reliability

It may be assumed that the overall reliability of a domestic clock is the lesser of the power source reliability and the movement reliability (repair or replacement). Using this definition the mean time between failures of a good quality mechanical clock is about 10 years, that of a good quality synchronous clock several months, and that of a quartz clock about 3 years.

6.3.3   Conclusions on Reliability

The implication is that it was the poor overall reliability that led to the fall of British domestic synchronous clock technology. Other shortcomings in some British domestic synchronous clocks were of secondary importance. The advent of quartz clocks merely hastened the fall. The AC mains supply is not under the control of British domestic synchronous clock manufacturers, so poor overall reliability could not be improved. British domestic synchronous clock manufacturers never took this into account. This was a serious marketing failure that had dramatic consequences.

6.4   The Rise and Fall of a Technology

The conclusions in Sects. 6.2.7 and 6.3.3 lead to a succinct explanation of the rise and fall of the technology. There were two booms in the sales of British domestic synchronous clocks. Despite their poor overall reliability, an initial boom in the 1930s was driven by the arrival in the UK of a must have new technology. This boom ended in 1939 with the outbreak of the Second World War, and the introduction of government restrictions on production. The second boom in the sales of British domestic synchronous clocks was driven by the pent up demand for domestic clocks following the end of the Second World War in 1945. In the 1970s pent up demand for domestic clocks had been satisfied. Hence, there was no longer an external driver for sales that overcame the poor overall reliability of British domestic synchronous clocks. The emergence of quartz clocks was a contributory factor, but not the main cause of the fall.

References

Anonymous (1935) New synchronous electric clock with carry-over mechanism. Horological J 77(917):131–132

Anonymous (1940a) Ten years of progress in making and selling synchronous electric clocks. Horological J 82(984):1–3

Anonymous (1940b) The future of the British clock manufacturing industry. B.H.I. meeting – official report. Horological J 82(978):96–100

Barrett DW (1939) Synchronous clocks and the jeweller. Horological J 81(968):75–76

Barrett DW (1946) Big future for British clocks. Official report of B.H.I. lecture. Horological J 88(1050):104–106

Bird C (2003) Metamec. The clockmaker. Dereham. Antiquarian Horological Society, Ticehurst

Davis L (2009) Ten years of anti-dumping in the EU: economic and political targeting. ECIPE Working Paper No. 02/2009 http.www.ecipe.org. Accessed 2014

Europa (2014) https://europa.eu. Accessed 2014

Lines MA (2012) Ferranti synchronous electric clocks, Paperback edition with corrections. Zazzo Media, Milton Keynes

Linz J (2004) Westclox electric. Schiffer Publishing Co., Atglen

Norman DA (2013) The design of everyday things. Revised and expanded edition. MIT Press, Cambridge, MA

Pook LP (2014) Temco Art Deco domestic synchronous clocks. Watch Clock Bull 56/1(407): 47–58

Read D (2012) The synchronous mains revolution and 'Time gentlemen, please'. Antiquarian Horology 33(4):487–492

Smith B (2008) Smiths domestic clocks, 2nd edn. Pierhead Publications Limited, Herne Bay

Wikipedia (2014) http://en.wikipedia.org. Accessed 2014

Part II
Galleries

Chapter 7
Gallery of Synchronous Mantel Clocks

Abstract A mantel clock is a clock intended for display either on a mantelpiece or on a piece of furniture. It has a dial that can be read from across the room. Domestic mantel clocks are mostly used in reception rooms. This is the commonest type of case used for synchronous clocks. The main purpose of this gallery is to illustrate the range of British synchronous mantel clocks that were available. A clock was assigned as a mantel clock primarily on the basis of the legibility of its dial. The dial on most of the clocks in this gallery is larger than 9 cm. There are back and front views of each clock, together with a brief description. Clocks are mostly illustrated as found, and some are in poor condition. The sample of 64 synchronous mantel clocks illustrated cannot be regarded as representative of clocks that were produced, or have survived. Nevertheless, the statistics provide a rough indication for clocks that have survived. The large number of Smith clocks (28 %) and Temco clocks (26 %) is noticeable. A number of clocks (23 %) have Art Deco cases, but there are few clocks with other identifiable artistic styles. Relatively few clocks (14 %) are self starting. None of them has an outage indicator. A significant number of clocks (30 %) do not have any form of tell tale to make it easy for a user to determine whether or not a clock is running. None of the clocks show the date, or the day of the week.

7.1 Introduction

A mantel clock is a clock intended for display either on a mantelpiece or on a piece of furniture. It has a dial that can be read from across the room. Domestic mantel clocks are mostly used in reception rooms. This is the commonest type of case used for synchronous clocks. The main purpose of this gallery is to illustrate the range of British synchronous mantel clocks that were available. There are back and front views of each clock, together with a brief description. Clocks are mostly illustrated as found, and some are in poor condition.

© Springer International Publishing Switzerland 2015   95
L.P. Pook, *British Domestic Synchronous Clocks 1930–1980*, History
of Mechanism and Machine Science 29, DOI 10.1007/978-3-319-14388-0_7

7.2   Alexander Clark Clock

Alexander Clark Synchronous Mantel Clock   See Figs. 4.5 and 7.1. This is a rebranded Smith *Mantel 05* clock, date range 1953–1959 (Smith 2008). The British Rail presentation plaque is undated. The clock is marked Made in Great Britain on the front, and Made in England on the back, with the door open. It is for a 200–250 V supply. The wood veneer case is 23 cm high × 30.5 cm wide. The dial is 15 cm diameter with a chrome bezel, black Roman numerals and gilt hands. The brand on the front is The Alexander Clark Co. Ltd, and the brand on the back, with the door open, is Smiths. The clock is fitted with a Smith self starting BM7 movement. There is no tell tale. British patent information on the clock is '366710, 369438, 374713'.

7.3   BEM Clock

BEM Synchronous Mantel Clock   See Fig. 7.2. The clock is marked Made in England. It is for a 200–250 V supply. The style is Art Deco. The gilt metal case is 12.5 cm high × 19.5 cm wide. The dial is 9 cm high × 13.5 cm wide with black Roman numerals and hands, and a red seconds hand. The brand on the front is BEM, and the brands on the back are BEM and British Electric Meters Ltd. The clock is fitted with a BEM self starting movement which has a Sangamo motor. The short seconds hand acts as a tell tale. British Patent information on the clock is '377696, 384880'.

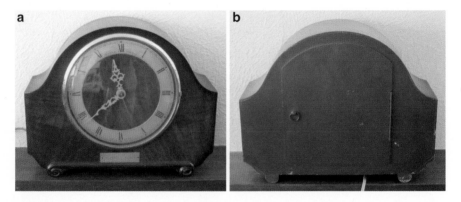

Fig. 7.1   Alexander Clark synchronous mantel clock (**a**) Front, (**b**) Back

Fig. 7.2 BEM synchronous mantel clock, (**a**) Front, (**b**) Back

Fig. 7.3 Camerer Kuss & Co synchronous mantel clock (**a**) Front, (**b**) Back

7.4 Camerer Kuss & Co Clock

Camerer Kuss and Co Synchronous Mantel Clock See Fig. 7.3. The clock is marked Made in Great Britain on the front, and Made in England on the back, with the door open. It is for a 200–250 V supply. The oak case is 34.5 cm high × 25 cm wide. The dial is 14.5 cm diameter with a chrome bezel, black Arabic numerals and hands, and a red seconds hand. The brands on the front are Smiths and SECTRIC. The brands on the back, with the door open, are Smiths English Clocks Ltd and

SEC. The brand on the base is Camerer, Kuss & Co. The clock is fitted with a Smith self starting Bijou movement. The seconds hand acts as a tell tale. British Patent information on the clock is '366710, 369438, 374713, 384441, 484222'.

7.5 Clyde Clocks

Clyde Synchronous Mantel Clock See Fig. 7.4. The clock is marked Made in Scotland. It is for a 200–250 V supply. The pale green plastic case is 14 cm high × 17 cm wide. The dial is 11.5 cm diameter with gilt Arabic numerals and hands. The brand on the back is Clyde. The clock is fitted with a Clyde movement. To start the clock push the starter lever and release. The rotor, just visible through the starter lever slot, acts as a tell tale.

Clyde Synchronous Mantel Clock See Fig. 7.5. The clock is marked British Made on the front and Made in Scotland on the back. It is for a 200–250 V supply. The oak veneer case is 14 cm high × 19.5 cm wide. The dial is 10 cm diameter with a gilt bezel, black Arabic numerals, and gilt hands. The clock is fitted with a Clyde movement. To start the clock push the starter lever and release. The rotor, just visible through the starter lever slot, acts as a tell tale.

7.6 Ediswan Clock

Ediswan Synchronous Mantel Clock See Fig. 7.6. The clock is marked Made in England. It is for a 200–250 V supply. The style is Geometric Octagonal. The wood veneer case is 19.5 cm high × 17 cm wide. The dial is 12.5 cm diameter with a

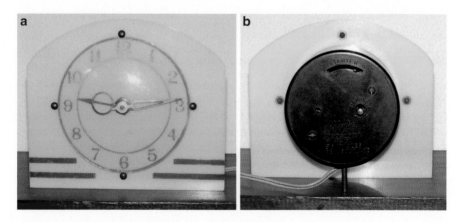

Fig. 7.4 Clyde synchronous mantel clock (**a**) Front, (**b**) Back

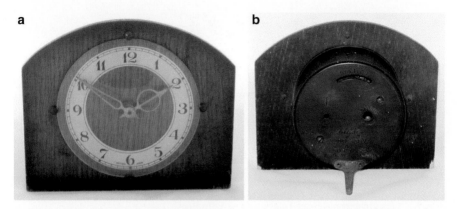

Fig. 7.5 Clyde synchronous mantel clock (**a**) Front, (**b**) Back

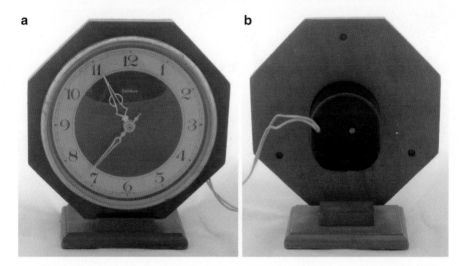

Fig. 7.6 Ediswan synchronous mantel clock (**a**) Front, (**b**) Back

chrome bezel, black Arabic numerals, and white hands. The brand on the front is Ediswan. The clock is fitted with an unbranded Smith Bijou movement. To start the clock press the starter knob and release. There is no tell tale.

7.7 Elco Clocks

Elco Synchronous Mantel Clock See Fig. 7.7. The clock is marked Made in England. It is for a 200–250 V supply. The style is Bakelite. The Bakelite case is 14 cm high × 20.5 cm wide. The dial is 9 cm diameter with black Arabic numerals

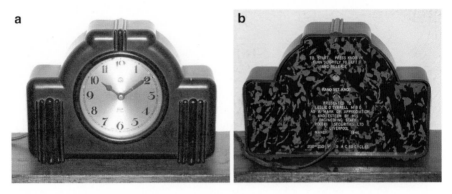

Fig. 7.7 Elco synchronous mantel clock (**a**) Front, (**b**) Back

Fig. 7.8 Elco synchronous mantel clock (**a**) Front, (**b**) Back

and hands. The brand on the front is Elco. The clock is fitted with an Elco movement. To start the clock press the starter knob and release. There is a tell tale in an aperture at 12 o'clock. The presentation marking on the back is dated March 1945. The back cover is made from Perspex, painted brown on the inside. The mottled effect is due to deterioration of the paint.

Elco Synchronous Mantel Clock See Fig. 7.8. The clock is marked Made in England on the front and Made in Great Britain on the back. It is for a 200–250 V supply. The wood veneer case is 20.5 cm high × 20.5 cm wide. The dial is 13.5 cm diameter with a chrome bezel, black Arabic numerals, chrome hands, and a black seconds hand. The brand on the back is Elco Clocks & Watches Ltd. The clock is fitted with an Elco movement. To start the clock press the starter knob and release. The seconds hand acts as a tell tale.

7.8 Empire Clock

Empire Synchronous Mantel Clock See Fig. 7.9. For more information see Miles (2011). The clock is marked Made in England. It is for a 200–250 V supply. The oak case is 16.5 cm high × 26 cm wide. The dial is 10 cm high × 14.5 cm wide with a chrome bezel, and black Arabic numerals and hands. The brand on the front and on the back is Empire. The clock is fitted with an Empire movement, serial number 10120. To start the clock push the starter lever and release. There is no tell tale. British Patent information on the clock is 'Patents pending'. The presentation plaque is dated 25 June 1935.

7.9 Ferranti Clocks

Model numbers and dates shown are based on information in Lines (2012). There are some inconsistencies. In particular, Ferranti sometimes used the same model number for different clocks produced at different times. It is not clear whether this was intentional or an oversight (Lines, personal communication 2012). There are often variations, for example in hand style, for clocks with the same model number produced in different years.

Ferranti Model No. 252 Synchronous Mantel Clock See Fig. 7.10. The date is 1938 (Lines 2012). The clock is marked Made in England. It is for a 200–250 V supply. The walnut veneer, ebonite and chrome case is 21.5 cm high × 21.5 cm wide. The dial is 19.5 cm diameter with an unglazed chapter ring, white Roman numerals, and gilt hands. The brand on the back is Ferranti. The clock is fitted with a Ferranti Early movement. To start the clock rotate the starter knob. The rotating starter knob acts as a tell tale. British Patent information on the clock is '379383, 384681'.

a   **b**

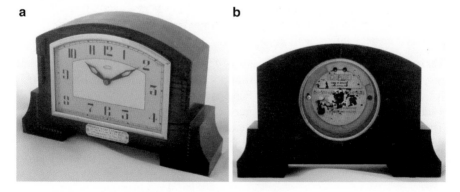

Fig. 7.9 Empire synchronous mantel clock (**a**) Front, (**b**) Back

Fig. 7.10 Ferranti Model No. 252 synchronous mantel clock (**a**) Front, (**b**) Back

Fig. 7.11 Ferranti Model No. 290 synchronous mantel clock (**a**) Front, (**b**) Back

Ferranti Model No. 290 Synchronous Mantel Clock See Fig. 7.11. The date range is 1947–1955 (Lines 2012). The clock is marked Made in England. It is for a 200–250 V supply. The clear Perspex and chrome case is 14.5 cm high × 14.5 cm wide. The dial is 11.5 cm diameter with black Arabic numerals and hands. The brand on the front and on the back is Ferranti. The clock is fitted with a Ferranti Later movement. To start the clock rotate the starter knob. The rotating starter knob acts as a tell tale. British Patent information on the clock is 'Patents pending'.

7.10   Garrard Clocks

Garrard Synchronous Mantel Clock See Fig. 7.12. The clock is marked Made in England. It is for a 200–250 V supply. The wood veneer case is 24 cm high × 26 cm wide. The dial is 17 cm square with a chrome bezel, and aluminium Arabic numerals and hands. The brand on the back is Garrard. The clock is fitted with a Garrard Early movement, serial number 5162. To start the clock twist the starter knob and release. There is a tell tale in an aperture in the movement cover. British Patent information on the clock is 'Patent pending'.

Garrard Synchronous Mantel Clock See Fig. 7.13. The clock is marked Made in England. It is for a 200–250 V supply. The walnut veneer case is 16.5 cm high × 26 cm wide. The dial is 11 cm high × 15 cm wide with black Roman numerals and white hands. The brand on the back is Garrard. The clock is fitted with a Garrard Early movement, serial number 8299. To start the clock twist the starter knob and release. There is no tell tale. British Patent information on the clock is 'Patent pending'.

Fig. 7.12 Garrard synchronous mantel clock (**a**) Front, (**b**) Back

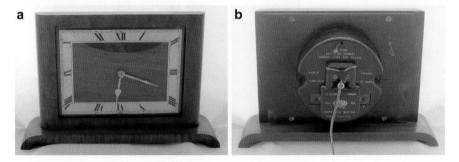

Fig. 7.13 Garrard synchronous mantel clock (**a**) Front, (**b**) Back

Fig. 7.14 Garrard synchronous mantel clock (**a**) Front, (**b**) Back

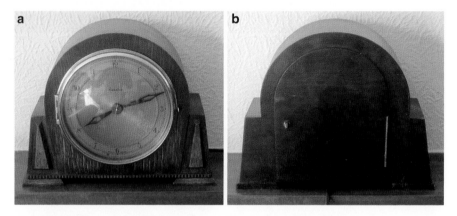

Fig. 7.15 Genalex striking synchronous mantel clock, (**a**) Front, (**b**) Back

Garrard Synchronous Mantel Clock See Fig. 7.14. The clock is marked Made in England. It is for a 200–250 V supply. The style is Art Deco. The black lacquered wood case is 18 cm high × 26.5 cm wide. The dial is 11.5 cm high × 14 cm wide with a chrome bezel, and black Roman numerals and hands. The brand on the front and on the back is Garrard. The clock is fitted with a Garrard Later movement, serial number 9403. To start the clock rotate the starter knob. The small seconds hand and the rotating starter knob act as tell tales. British Patent information on the clock is 'Patent pending'.

7.11   Genalex Clocks

Genalex Striking Synchronous Mantel Clock See Fig. 7.15. For more information see Miles (2011). The clock is marked Made in England. It is for a 200–250 V supply. The wood veneer case is 19.5 cm high × 25.5 cm wide. The dial is 12 cm

Fig. 7.16 Genalex synchronous mantel clock, (**a**) Front, (**b**) Back

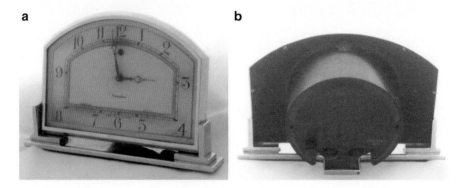

Fig. 7.17 Genalex synchronous mantel clock, (**a**) Front, (**b**) Back

diameter with a chrome bezel, and black Arabic numerals and hands. The brand on the front and on the back, with the door open, is Genalex. The clock is fitted with a Smith Type 1 striking movement. To start the clock push the starter lever and release. There is a tell tale in an aperture on the back plate.

Genalex Synchronous Mantel Clock See Fig. 7.16. This is a rebranded Smith *Carisbrook*, date range 1932–1936 (Smith 2008). The clock is marked Made in England. It is for a 200–250 V supply. The style is Art Deco. The wood veneer and chrome case is 23 cm high × 27.5 cm wide. The dial is 17 cm high × 21 cm wide with black Roman numerals and brass hands. The brand on the back is Genalex. The clock is fitted with a Smith BM7 movement. To start the clock press the starter knob and release. There is no tell tale. British Patent information on the clock is '366710, 369438, 374713'.

Genalex Synchronous Mantel Clock See Fig. 7.17. This is a rebranded Smith *Ceylon*, date range 1935–1936 (Smith 2008), but with different hands. The clock is marked Made in England. It is for a 200–250 V supply. The chrome case is

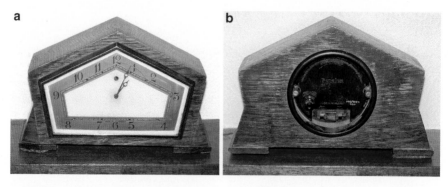

Fig. 7.18 Genalex synchronous mantel clock, (**a**) Front, (**b**) Back

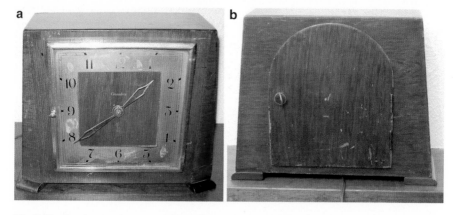

Fig. 7.19 Genalex synchronous striking mantel clock, (**a**), Front, (**b**) Back

14.5 cm high × 18 cm wide. The dial is 9 cm high × 13.5 cm wide with black Arabic numerals and hands. The brand on the front and on the back is Genalex. The clock is fitted with a Smith BM7 movement. To start the clock press the starter knob and release. There is a tell tale in an aperture at 12 o'clock. British Patent information on the clock is '366710, 369438, 374713, others pending'.

Genalex Synchronous Mantel Clock See Fig. 7.18. The clock is marked Made in England. It is for a 200–250 V supply. The oak case is 15 cm high × 24 cm wide. The dial is 10 cm high × 21 cm wide with a chrome bezel, and black Arabic numerals and hands. The clock is fitted with a Smith BM7 movement. To start the clock press the starter knob and release. There is a tell tale in an aperture at 12 o'clock. British Patent information on the clock is '366710, 369438, 384441, others pending'.

Genalex Synchronous Striking Mantel Clock See Fig. 7.19. The clock is marked Made in England. It is for a 200–250 V supply. The wood veneer case is 18.5 cm high × 23.5 cm wide. The dial is 12.5 cm square with a chrome bezel, black Arabic numerals, and chrome hands. The brand on the front is Genalex. The brand on the

Fig. 7.20 Ingersoll synchronous mantel clock, (**a**) Front, (**b**) Back

back, with the door open, is Smiths English Clocks Ltd. The clock is fitted with a Smith Thin striking movement. To start the clock push the starter lever and release. There is a tell tale in an aperture in the back plate. British patent information on the clock is '387108'.

7.12 Ingersoll Clocks

Ingersoll Synchronous Mantel Clock See Fig. 7.20. The clock is marked Made in England. It is for a 200–250 V supply. The wood veneer and chrome case is 12.5 cm high × 23 cm wide. The dial is 10 cm diameter with a chrome bezel, and brown Arabic numerals and hands. The brand on the front and on the back is Ingersoll. The clock is fitted with a Smith Bijou movement branded Ingersoll. To start the clock press the starter knob and release. There is no tell tale. British Patent information on the clock is '366710, 369438, 374713, others pending'.

Ingersoll Synchronous Mantel Clock See Fig. 7.21. The clock is marked Made in England. It is for a 200–250 V supply. The wood veneer case is 20.5 cm high × 23 cm wide. The dial is 14 cm diameter with a brassed bezel, and brass Arabic numerals and hands. The brand on the front and on the back is Ingersoll. The clock is fitted with a Smith Bijou movement. To start the clock press the starter knob and release. There is no tell tale. British Patent information on the clock is '366710, 369438, 374713, 384441, 484222'.

7.13 Liberty Clock

Liberty Synchronous Mantel Clock See Fig. 7.22. The clock is marked Made in England. It is for a 200–250 V supply. The style is G-Plan. The clock was probably made in the 1950s. The marble case is 19.5 cm high × 33.5 cm wide. The dial is

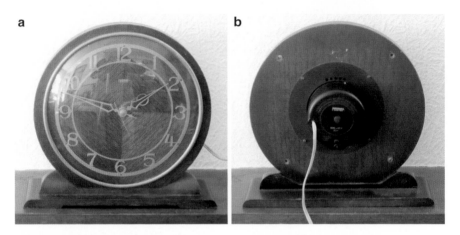

Fig. 7.21   Ingersoll synchronous mantel clock, (**a**) Front, (**b**) Back

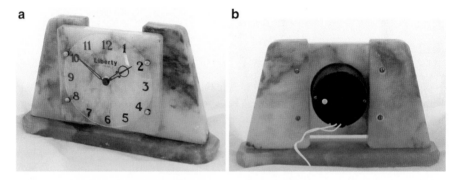

Fig. 7.22   Liberty synchronous mantel clock, (**a**) Front, (**b**) Back

15 cm diameter with black Arabic numerals and hands. The brand on the front is Liberty. The brands on the back are Temco and Telephone Mfg. Co. Ltd. The clock is fitted with a Temco Mark V movement. To start the clock rotate the starter knob. The rotating starter knob acts as a tell tale.

7.14   Marigold Clock

Marigold Synchronous Mantel Clock   See Fig. 7.23. The clock is marked Made in England. It is for a 200–250 V supply. The style is Art Deco. The oak case is 15 cm high × 19 cm wide. The dial is 10 cm square with a chrome bezel, and black Arabic numerals and hands. The brand on the front is Marigold. The clock is fitted

Fig. 7.23 Marigold synchronous mantel clock, (**a**) Front, (**b**) Back

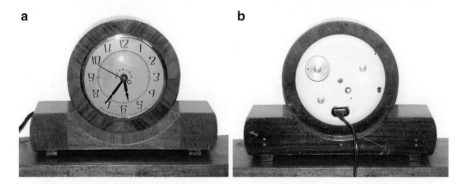

Fig. 7.24 Metalair synchronous mantel clock, (**a**) Front, (**b**) Back

with a Marigold movement. To start the clock push the starter lever and release. There is a tell tale in an aperture at 6 o'clock. British Patent information on the clock is '444688'.

7.15   Metalair Clock

Metalair Synchronous Mantel Clock   See Fig. 7.24. The clock is marked Manufactured in England. It is for a 200–250 V supply. The style is Art Deco. The wood veneer case is 19 cm high × 25.5 cm wide. The dial is 10 cm diameter with a chrome bezel, and black Arabic numerals and hands. The brands on the front are Metalair and Metalair Ltd. The clock is fitted with a Metalair movement. To start the clock twist the starter knob and release. The seconds hand acts as a tell tale.

7.16   Metamec Clocks

Metamec Model 945 Synchronous Mantel Clock See Fig. 7.25. For more information see Bird (2003). The clock is marked Made in England. It is for a 220–240 V supply. The rosewood case is 14 cm high × 18.5 cm wide. The dial is 11 cm diameter with a brass bezel, and gilt Arabic numerals and hands. The brand on the front and on the back is Metamec. The clock is fitted with a Metamec self starting SS 101 movement. The seconds hand acts as a tell tale.

Metamec Model 946 Synchronous Mantel Clock See Fig. 7.26. The date range is September 1967–August 1993 (Bird 2003). The clock is marked Made in England. It is for a 220–240 V supply. The wood and brass case is 14.5 cm high × 25.5 cm wide. The dial is 10 cm diameter with a brass bezel, black Roman numerals and hands, and a gilt seconds hand. The brand on the front and on the back is Metamec. The clock is fitted with a Metamec self starting SS 101 movement. The seconds hand acts as a tell tale.

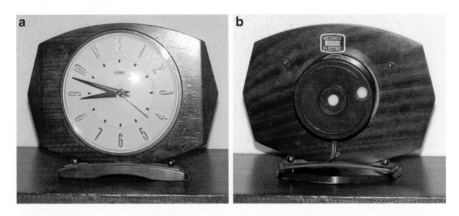

Fig. 7.25 Metamec Model 945 synchronous mantel clock, (**a**) Front, (**b**) Back

Fig. 7.26 Metamec Model 946 synchronous mantel clock, (**a**) Front, (**b**) Back

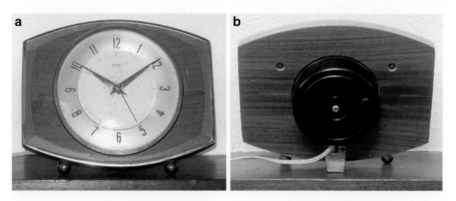

Fig. 7.27   Metamec synchronous mantel clock, (**a**) Front, (**b**) Back

Fig. 7.28   Riley synchronous mantel clock, (**a**) Front, (**b**) Back

Metamec Synchronous Mantel Clock See Fig. 7.27. This is the synchronous version of the Metamec Model B 26 8 day mechanical mantel clock (Bird 2003). The clock is marked Made in England. It is for a 200–250 V supply. The wood veneer and brass case is 14 cm high × 18 cm wide. The dial is 10 cm diameter with a brassed bezel, black Arabic numerals, and brass hands. The brand on the front and on the back is Metamec. The clock is fitted with a Metamec self starting SS 101 movement. The seconds hand acts as a tell tale.

7.17   Riley Clock

Riley Synchronous Mantel Clock See Fig. 7.28. The clock is marked Made in England. It is for a 200–250 V supply. The style is Geometric Hexagonal. The wood veneer case is 18 × 19 cm wide. The dial is 14 cm diameter with a chrome bezel,

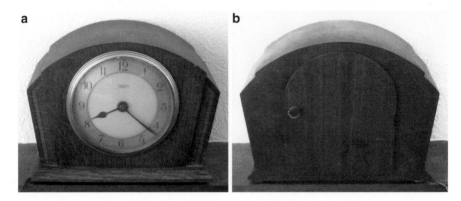

Fig. 7.29 Smith *Sudan* synchronous striking mantel clock, (**a**) Front, (**b**) Back

black Arabic numerals, and gilt hands. The brand on the front is Riley. The clock is fitted with a Smith Bijou movement. To start the clock twist the starter knob and release. There is no tell tale. British Patent information on the clock is 'Patent pending'.

7.18   Smith Clock, Type 1 Striking Movement

Smith *Sudan* Synchronous Striking Mantel Clock   See Fig. 7.29. The date is 1934 (Smith 2008). The clock is marked Made in England. It is for a 200–250 V supply. The wood veneer case is 20.5 cm high × 26.5 cm wide. The dial is 13.5 cm diameter with a chrome bezel, and black Arabic numerals and hands. The brand on the front is Smith. The brands on the back are Smiths English Clocks Ltd and SEC. The clock is fitted with a Smith Type 1 striking movement, serial number 6153. To start the clock push the starter lever and release. There is a tell tale in an aperture in the back plate. British Patent information on the clock is '366710, others pending'.

7.19   Smith Clock, Westminster Chiming Movement

Smith *Kendall* Chiming Mantel Clock   See Fig. 7.30. The date range is 1953–1955 (Smith 2008). The clock is marked Made in Great Britain on the front and Made in England on the back, with the door open. It is for a 200–250 V supply. The oak veneer case is 23 cm high × 28 cm wide. The dial is 16 cm diameter with a brass bezel, black Roman numerals, and brass hands. The brands on the front are Smith and SECTRIC. The clock is fitted with a Smith self starting Westminster chiming movement. There is no tell tale. British Patent information on the clock is '366710, 387108, 412336, others pending'.

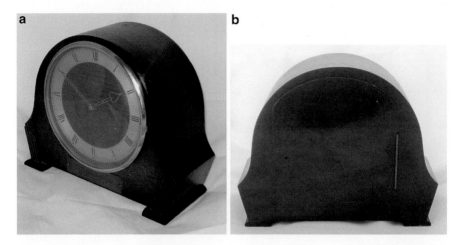

Fig. 7.30 Smith *Kendall* chiming mantel clock, (**a**) Front, (**b**) Back

Fig. 7.31 Smith *Chalfont* synchronous mantel clock, (**a**) Front, (**b**) Back

7.20   Smith Clocks, BM7 Movement

Smith *Chalfont* Synchronous Mantel Clock   See Fig. 7.31. The date range is 1932–1939 (Smith 2008). The clock is marked Made in England. It is for a 200–250 V supply. The wood veneer case is 21.5 cm high × 24 cm wide. The dial is 12 cm high × 16 cm wide with a chrome bezel, black Roman numerals, and white hands. The brand on the front is Smith. Brands on the back are SEC, Smith's English Clocks Ltd, and Synchronous Electric Clocks Ltd. The clock is fitted with a Smith BM7 movement. To start the clock press the starter knob and release. There is a tell tale in an aperture at 12 o'clock. British Patent information on the clock is '366710, 369438, 374713, others pending'.

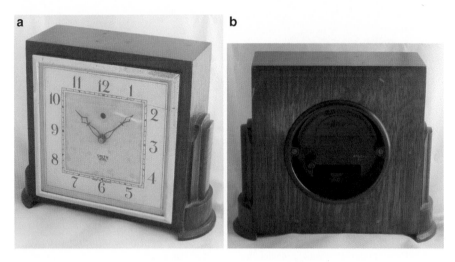

Fig. 7.32 Smith *Tees* synchronous mantel clock, (**a**) Front, (**b**) Back

Fig. 7.33 Smith synchronous mantel clock, (**a**) Front, (**b**) Back

Smith *Tees* Synchronous Mantel Clock See Fig. 7.32. The date is 1937 (Smith 2008). The clock is marked Made in England. It is for a 200–250 V supply. The oak case is 16.5 cm high × 21.5 cm wide. The dial is 12.5 cm square with a chrome bezel, and black Arabic numerals and hands. Brands on the front are Smith and SECTRIC. Brands on the back are SEC, Smith's English Clocks Ltd, and Synchronous Electric Clocks Ltd. The clock is fitted with a Smith BM7 movement. To start the clock press the starter knob and release. There is a tell tale in an aperture at 12 o'clock. British Patent information on the clock is '366710, 369438, 374713, others pending'.

Smith Synchronous Mantel Clock See Fig. 7.33. The clock is marked Made in England. It is for a 200–250 V supply. The oak case is 18 cm high × 21.5 cm

Fig. 7.34 Smith synchronous mantel clock, (**a**) Front, (**b**) Back

wide. The dial is 12.5 cm square with a chrome bezel, and black Arabic numerals and hands. Brands on the front are Smith and SECTRIC. Brands on the back are SEC, Smith's English Clocks Ltd and Synchronous Electric Clocks Ltd. The clock is fitted with a Smith BM7 movement. To start the clock press the starter knob and release. There is a tell tale in an aperture at 12 o'clock. British Patent information on the clock is '366710, 369438, 374713, others pending'.

Smith Synchronous Mantel Clock See Fig. 7.34. The clock is marked Made in England. It is for a 200–250 V supply. The oak case is 18 cm high × 23 cm wide. The dial is 15 cm square with a chrome bezel, and black Arabic numerals and hands. Brands on the front are Smith and SECTRIC. Brands on the back are SEC, Smith's English Clocks Ltd and Synchronous Electric Clocks Ltd. The clock is fitted with a Smith BM7 movement. To start the clock press the starter knob and release. There is a tell tale in an aperture at 12 o'clock. British Patent information on the clock is '366710, 369438, 374713, others pending'.

Smith Synchronous Mantel Clock See Figs. 4.4 and 7.35. The clock is marked Made in England. It is for a 200–250 V supply. The style is Chinoiserie. The lacquered wood case is 16 cm high × 24 cm wide. The dial is 10 cm square with a chrome bezel, and black Arabic numerals and hands. Brands on the front are Smith and SECTRIC. Brands on the back are SEC, Smith's English Clocks Ltd and Synchronous Electric Clocks Ltd. The clock is fitted with a Smith BM7 movement. The front plate is marked 'BM7/36/8'. '36/8' is probably the date of manufacture, August 1936. To start the clock press the starter knob and release. There is a tell tale in an aperture at 12 o'clock. British Patent information on the clock is '366710, 369438, 374713'.

Smith Synchronous Mantel Clock See Figs. 4.6 and 7.36. The clock is marked Made in England. It is for a 200–250 V supply. The style is Egyptian. The embellished marble case is 20.5 cm high × 35.5 cm wide. The garnitures are 14.5 cm high × 11 cm wide. The dial is 8.5 cm diameter with a chrome bezel, and black

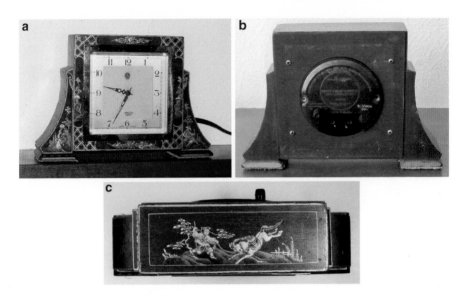

Fig. 7.35 Smith synchronous mantel clock, (**a**) Front, (**b**) Back, (**c**) Top

Fig. 7.36 Smith synchronous mantel clock, (**a**) Front, (**b**) Back

Arabic numerals and hands. It is an imposing clock, although the dial is a little too small for a mantel clock. The Egyptian style animals suggest that the clock was made to commemorate a Tutankhamen exhibition, probably the one held in 1952. Brands on the front are Smith and SECTRIC. Brands on the back are SEC, Smith's English Clocks Ltd and Synchronous Electric Clocks Ltd. The clock is fitted with a Smith BM7 movement. To start the clock press the starter knob and release. There is a tell tale in an aperture at 12 o'clock. British Patent information on the clock is '366710, 369438, 374713, others pending'.

7.21 Smith Clocks, Bijou Movement

Smith *Essex* Synchronous Mantel Clock See Fig. 7.37. The date is 1948 (Smith 2008). The clock is marked Made in England. It is for a 200–250 V supply. The style is Art Deco. The cream and black plastic case is 17 cm high × 19.5 cm wide. The dial is 15 cm diameter with a chrome bezel, brown Arabic numerals and batons, and black hands. Brands on the front are Smith and SECTRIC. Brands on the back are SEC and Smiths English Clocks. The clock is fitted with a Smith Bijou movement. The front plate is marked '15 8 47'. This appears to be the date of manufacture, 15 August 1947. To start the clock press the starter knob and release. There is no tell tale. British Patent Information on the clock is '366710, 369438, 374713, 384441, 484222'.

Smith *Grenfell* Synchronous Mantel Clock See Fig. 7.38. The date range is 1954–1958 (Smith 2008). The clock is marked Made in England. It is for a 200–250 V supply. The brass and wood case is 23 cm high × 29 cm wide. The dial is 17 cm diameter with brass Arabic numerals and hands. Brands on the front are Smiths and

Fig. 7.37 Smith *Essex* synchronous mantel clock, (**a**) Front, (**b**) Back

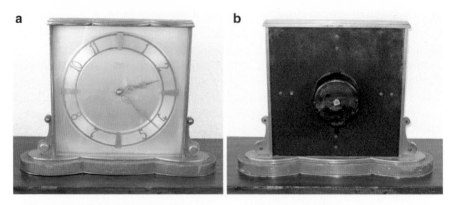

Fig. 7.38   Smith *Grenfell* synchronous mantel clock, (**a**) Front, (**b**) Back

Fig. 7.39   Smith *Tapton* synchronous mantel clock, (**a**) Front, (**b**) Back

SECTRIC. Brands on the back are Smiths English Clocks and SEC. The clock is fitted with a Smith Bijou movement. To start the clock press the starter knob and release. There is no tell tale. British Patent Information on the clock is '366710, 369438, 374713, 384441, 484222'.

Smith *Tapton* Synchronous Mantel Clock See Fig. 7.39. The date is 1939 (Smith 2008). The clock is marked Made in England. It is for a 200–250 V supply. The wood veneer case is 21.5 cm high × 19 cm wide. The dial is 12.5 cm square with a chrome bezel, and black Roman numerals and hands. The dial centre is beige fabric. Brands on the back are Smiths English Clocks and SEC. The clock is fitted with a Smith Bijou movement. To start the clock press the starter knob and release. There is no tell tale. British Patent information on the clock is '366710, 369438, 374713, 384441, 484222'.

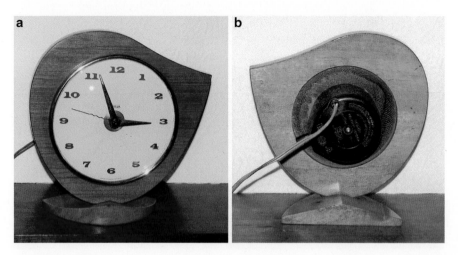

Fig. 7.40 Smith *Woburn* variant synchronous mantel clock, (**a**) Front, (**b**) Back

Smith *Woburn* Variant Synchronous Mantel Clock See Figs. 4.8 and 7.40. The case is the same as that of the Smith Woburn, date 1961, but the dial is the same as that of the Smith *Yvette*, date range date range 1963–1965 (Smith 2008). The clock is marked Made in Great Britain. It is for a 200–250 V supply. The style is G-Plan. The wood case is 18.5 cm high × 18 cm wide. The dial is 11.5 cm diameter with a gilt bezel, and gilt Arabic numerals and hands. The brand on the front is Smiths. Brands on the back are Smiths English Clocks and SEC. The clock is fitted with a Smith Bijou movement. To start the clock press the starter knob and release. The seconds hand acts as a tell tale.

Smith Synchronous Mantel Clock See Fig. 7.41. The clock is marked Made in England. It is for a 200–250 V supply. The oak case is 18 cm high × 19.5 cm wide. The dial is 12 cm square with a chrome bezel, black Roman numerals, and gilt hands. The brands on the front are Smith and SECTRIC. The brands on the back are SEC and Smiths English Clocks. The clock is fitted with a Smith Bijou movement. To start the clock press the starter knob and release. There is no tell tale. British Patent information on the clock is '366710, 369438, 374713, 384441, 484222'.

Smith Synchronous Mantel Clock See Fig. 7.42. The clock is marked Made in England. It is for a 200–250 V supply. The oak case is 18.5 cm high × 21.5 cm wide. The dial is 13.5 cm diameter with a chrome bezel, black Roman numerals, and white hands. Brands on the front are Smiths and SECTRIC. Brands on the back are SEC and Smiths English Clocks. The clock is fitted with a Smith Bijou movement. To start the clock press the starter knob and release. There is no tell tale. British Patent information on the clock is '366710, 369438, 374713, 384441, 484222'.

Smith Synchronous Mantel Clock See Fig. 7.43. The clock is marked Made in England. The style is Chinoiserie. The lacquered wood case is 20.5 cm high × 15 cm wide. The dial is 12.5 cm diameter with a bronze bezel, black Roman numerals, and

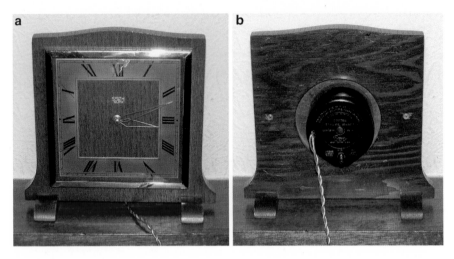

Fig. 7.41 Smith synchronous mantel clock, (**a**) Front, (**b**) Back

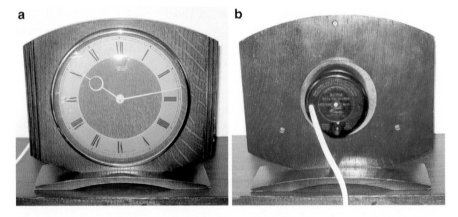

Fig. 7.42 Smith synchronous mantel clock, (**a**) Front, (**b**) Back

aluminium hands. Brands on the back are Smiths English Clocks and SEC. The clock is fitted with a Smith Bijou movement. To start the clock press the starter knob and release. There is no tell tale. British Patent information on the clock is '366710, 374713, 384441, 484222'.

7.22   Smith Clock, QAT Movement

Smith *Huntingdon* Synchronous Mantel Clock See Figs. 4.10 and 7.44. The date is 1953 (Smith 2008). The clock is marked Made in England. It is for a 200–250 V supply. The styles are Pictorial and Bakelite. The Bakelite case is

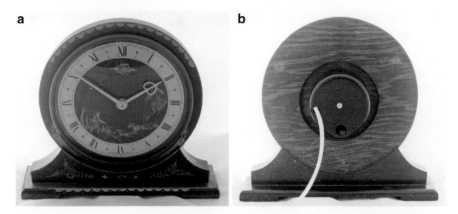

Fig. 7.43 Smith synchronous mantel clock, (**a**) Front, (**b**) Back

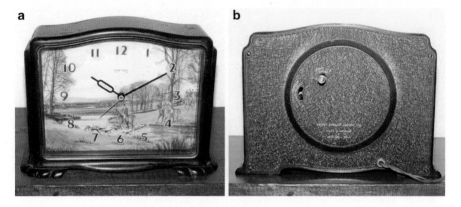

Fig. 7.44 Smith *Huntingdon* synchronous mantel clock, (**a**) Front, (**b**) Back

14 cm high × 19.5 cm wide. The dial is 11.5 cm high × 14 cm wide with black Arabic numerals and hands, and a brass seconds hand. The brand on the front is Smiths. The brand on the back is Smiths English Clocks Ltd. The clock is fitted with a Smith QAT movement. To start the clock press the starter knob and release. The seconds hand acts as a tell tale.

7.23   Smith Clocks, QEMG Movement

Smith *Radbourne* Synchronous Mantel Clock   See Fig. 7.45. The date range is 1950–1956 (Smith 2008). The clock is marked Made in Great Britain on the front and Made in England on the back. It is for a 200–250 V supply. The style is Bakelite. The Bakelite case is 14 cm high × 19.5 cm wide. The dial is 10 cm diameter with

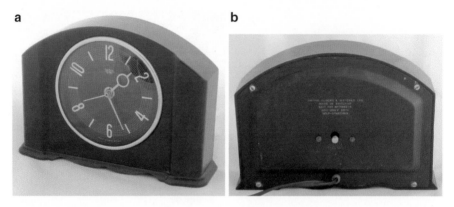

Fig. 7.45 Smith *Radbourne* synchronous mantel clock, (**a**) Front, (**b**) Back

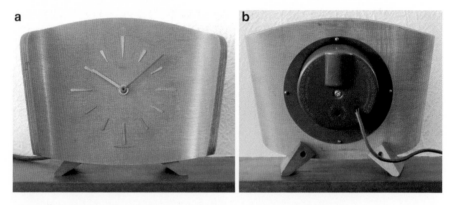

Fig. 7.46 Smith *Woodhaven* synchronous mantel clock, (**a**) Front, (**b**) Back

a chrome bezel, white Arabic numerals, batons and hands, and a brass seconds hand. Brands on the front are Smith and SECTRIC. The brand on the back is Smiths Clocks & Watches Ltd. The clock is fitted with a Smith self starting QEMG movement. The seconds hand acts as a tell tale. British Patent information on the clock is '744204'.

Smith *Woodhaven* Synchronous Mantel Clock See Fig. 7.46. The date range is 1957–1958 (Smith 2008). The clock is marked Made in Great Britain on the front and Made in England on the back. It is for a 200–250 V supply. The style is G-Plan. Brands on the front are Smith and SECTRIC. The brand on the back is Smiths Clocks & Watches Ltd. The clock is fitted with a Smith self starting QEMG movement. The movement is marked 'GEMG 3101 587': '587' indicates the date of manufacture, July 1958. There is no tell tale. British Patent information on the clock is '744204, others pending'.

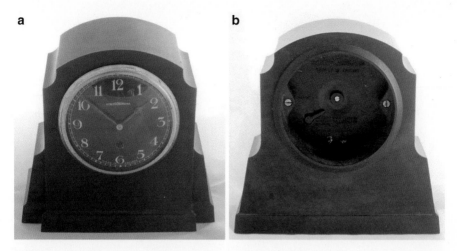

Fig. 7.47 Synchronomains synchronous mantel clock, (**a**) Front, (**b**) Back

7.24   Synchronomains Clock

Synchronomains Synchronous Mantel Clock   See Fig. 7.47. The clock is marked Made in England. The style is Bakelite. The Bakelite case is 16.5 cm high × 17 cm wide. The dial is 9.5 cm diameter with a chrome bezel, white Arabic numerals, and blued steel hands. The brand on the front and on the back is Synchronomains. The clock is fitted with an Empire movement, serial number 7310. To start the clock push the starter lever and release. There is a tell tale in an aperture at 12 o'clock. British Patent information on the clock is 'Patent pending'.

7.25   Temco Clocks, Mark I Movement

Temco Synchronous Mantel Clock   See Figs. 4.1c and 7.48. For more information see Pook (2014). The clock is marked Made in England. It is for a 200–240 V supply. The style is Art Deco. The Bakelite case with brass decoration is 24 cm high × 20.5 cm wide. The dial is 12.5 cm diameter with a brass bezel, black Roman numerals, and brass hands. There is a seconds indicator in an aperture at 5 o'clock. The brand on the back is Telephone Mfg. Co. Ltd. The clock is fitted with a Temco Mark I movement. To start the clock press the starter knob and release. The seconds indicator acts as a tell tale. British Patent information on the clock is 'Patents pending'.

Temco Synchronous Mantel Clock   See Fig. 7.49. The clock is marked Made in England. It is for a 200–240 V supply. The style is Art Deco. The Bakelite case is 20.5 cm high × 20.5 cm wide. The dial is 12 cm diameter with black Roman

Fig. 7.48   Temco synchronous mantel clock, (**a**) Front, (**b**) Back

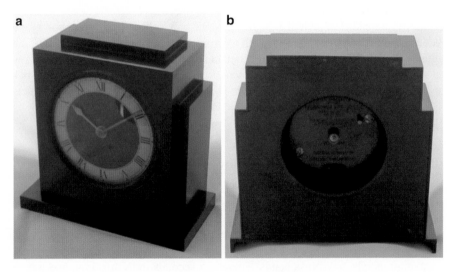

Fig. 7.49   Temco synchronous mantel clock, (**a**) Front, (**b**) Back

numerals and brass hands. The brand on the back is Telephone Mfg. Co. Ltd. The clock is fitted with a Temco Mark I movement. To start the clock press the starter knob and release. The seconds indicator acts as a tell tale. British Patent information on the clock is 'Patents pending'.

7.26 Temco Clocks, Mark IV Movement

Temco Synchronous Mantel Clock See Fig. 7.50. For more information see Pook (2014). The clock is marked Made in England. It is for a 200–250 V supply. The style is Art Deco. The oak veneer case is 21.5 cm high × 27.5 cm wide. The dial is 13.5 cm square with a chrome bezel, black Roman numerals, and chrome hands. Brands on the back are Temco and Telephone Mfg. Co. Ltd. The clock is fitted with a Temco Mark IV movement, serial number 23536. To start the clock rotate the starter knob. The rotating starter knob acts as a tell tale.

Temco Synchronous Mantel Clock See Figs. 4.1a and 7.51. The clock is marked Made in England. It is for a 200–250 V supply. The style is Art Deco. The wood veneer case is 17 cm high × 28 cm wide. The unglazed dial is 12.5 cm high × 15 cm wide with white Roman numerals and hands. Brands on the back are Temco and Telephone Mfg. Co. Ltd. The clock is fitted with a Temco Mark IV movement,

Fig. 7.50 Temco synchronous mantel clock, (**a**) Front, (**b**) Back

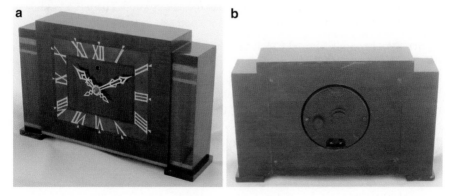

Fig. 7.51 Temco synchronous mantel clock (**a**) Front, (**b**) Back

Fig. 7.52 Temco synchronous mantel clock, (**a**) Front, (**b**) Back

Fig. 7.53 Temco synchronous mantel clock, (**a**) Front, (**b**) Back

serial number 28484. To start the clock rotate the starter knob. There is a tell tale in an aperture at 12 o'clock. The rotating starter knob acts as a tell tale.

Temco Synchronous Mantel Clock See Fig. 7.52. For more information see Pook (2014). The clock is marked Made in England. It is for a 200–250 V supply. The style is Art Deco. The oak case is 20.5 cm high × 28.5 cm wide. The unglazed dial is 19.5 cm high × 23.5 cm wide with brass Arabic numerals and hands. Brands on the back are Temco and Telephone Mfg. Co. Ltd. The clock is fitted with a Temco Mark IV movement, serial number 28956. To start the clock rotate the starter knob. There is a tell tale in an aperture at 12 o'clock. The rotating starter knob acts as a tell tale.

Temco Synchronous Mantel Clock See Figs. 4.2b and 7.53. For more information see Pook (2014). The clock is marked Made in England. It is for a 200–250 V supply.

The style is Art Deco. The Bakelite case is 20.5 cm high × 20.5 cm wide. The dial is 12.5 cm diameter with black Roman numerals, and brass hands. Brands on the back are Temco and Telephone Mfg. Co. Ltd. The clock is fitted with a Temco Mark IV movement. To start the clock rotate the starter knob. There is a tell tale in an aperture at 12 o'clock, and the rotating starter knob acts as a tell tale.

7.27   Temco Clocks, Mark V Movement

Temco No. 1200 Synchronous Mantel Clock   See Figs. 4.12 and 7.54. The clock was advertised in 1940 Anonymous (1940). The clock is marked Made in England. It is for a 200–250 V supply. The style is Wrought Iron. The black and gilt wrought iron case is 30.5 cm high × 34.5 cm wide. The unglazed dial is 18 cm diameter with gilt Roman numerals and hands. Brands on the back are Temco and Telephone Mfg. Co. Ltd. The clock is fitted with a Temco Mark V movement. To start the clock rotate the starter knob. The rotating starter knob acts as a tell tale.

Temco No. 4750 Synchronous Mantel Clock   See Fig. 7.55. For more information see Anonymous (1948). The clock is marked Made in England. It is for a 200–250 V supply. The clear Perspex, orange plastic, aluminium, brass and black Bakelite case is 14 cm high × 23 cm wide. The designer clearly intended that the black Bakelite movement cover would be visible as part of the case. The dial is 11.5 cm diameter with pale yellow Arabic numerals and batons, and white hands. The numerals, batons and circles are printed on the back of the clear Perspex. The grey background to the dial is matt finish aluminium behind the clear Perspex. The brand on the front is Temco. Brands on the back are Temco and Telephone Mfg. Co. Ltd. The clock is fitted with a Temco Mark V movement. To start the clock rotate the starter knob. The rotating starter knob acts as a tell tale.

Fig. 7.54 Temco No. 1200 synchronous mantel clock, (**a**) Front, (**b**) Back

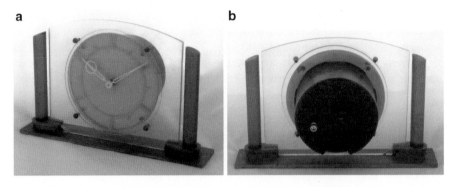

Fig. 7.55 Temco No. 4750 synchronous mantel clock, (**a**) Front, (**b**) Back

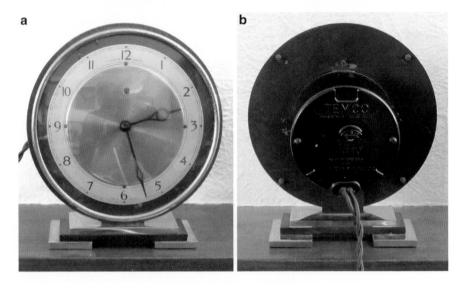

Fig. 7.56 Temco synchronous mantel clock, (**a**) Front, (**b**) Back

Temco Synchronous Mantel Clock See Fig. 7.56. The clock is marked Made in England. It is for a 200–250 V supply. The style is Art Deco. The chrome and stained wood case is 19 cm high × 17 cm wide. The dial is 12.5 cm diameter with a chrome bezel, and black Arabic numerals and hands. The brand on the front is Temco. Brands on the back are Temco and Telephone Mfg. Co. Ltd. The clock is fitted with a Temco Mark V movement. To start the clock rotate the starter knob. There is a tell tale in an aperture at 12 o'clock. The rotating starter knob acts as a tell tale.

Temco Synchronous Mantel Clock See Fig. 7.57. For more information see Pook (2014). The clock is marked Made in England. It is for a 200–250 V supply. The veneered plywood case is 16.5 cm high × 23 cm wide. The dial is 12 cm square with a chrome bezel, black Roman numerals, and white hands. The brand on the front is

Fig. 7.57 Temco synchronous mantel clock, (**a**) Front, (**b**) Back

Fig. 7.58 Temco synchronous mantel clock, (**a**) Front, (**b**) Back

Temco. Brands on the back are Temco and Telephone Mfg. Co. Ltd. The clock is fitted with a Temco Mark V movement. To start the clock rotate the starter knob. There is a tell tale in an aperture at 12 o'clock. The rotating starter knob acts as a tell tale.

Temco Synchronous Mantel Clock See Fig. 7.58. The clock is marked Made in England. It is for a 200–250 V supply. The veneered plywood case is 15 cm high × 20.5 cm wide. The dial is 12 cm square with a chrome bezel, and black Arabic numerals and hands. The brand on the front is Temco. Brands on the back are Temco and Telephone Mfg. Co. Ltd. The clock is fitted with a Temco Mark V movement. To start the clock rotate the starter knob. There is a tell tale in an aperture at 12 o'clock. The rotating starter knob acts as a tell tale.

Temco Synchronous Mantel Clock See Fig. 7.59. The clock is marked Made in England. It is for a 200–250 V supply. The oak case is 14 cm high × 16.5 cm wide. The dial is 11 cm square with a chrome bezel, and black Arabic numerals and hands.

Fig. 7.59 Temco synchronous mantel clock, (**a**) Front, (**b**) Back

Fig. 7.60 Temco synchronous mantel clock, (**a**) Front, (**b**) Back

The brand on the front is Temco. Brands on the back are Temco and Telephone Mfg. Co. Ltd. The clock is fitted with a Temco Mark V movement. To start the clock rotate the starter knob. There is a tell tale in an aperture at 12 o'clock. The rotating starter knob acts as a tell tale.

Temco Synchronous Mantel Clock See Fig. 7.60. The clock is marked Made in England. It is for a 200–250 V supply. The walnut veneer case is 23 cm high × 26.5 cm wide. The dial is 12.5 cm diameter with a chrome bezel, a cream skeletonised chapter ring with Arabic numerals, and cream hands. The chapter ring is not quite circular and four of the numerals are cut outs. Brands on the back are Temco and Telephone Mfg. Co. Ltd. The clock is fitted with a Temco Mark V movement. To start the clock rotate the starter knob. There is a tell tale in an aperture at 12 o'clock. The rotating starter knob acts as a tell tale.

Temco Synchronous Mantel Clock See Fig. 7.61. The clock is marked Made in England. It is for a 200–250 V supply. The oak veneer case is 20.5 cm

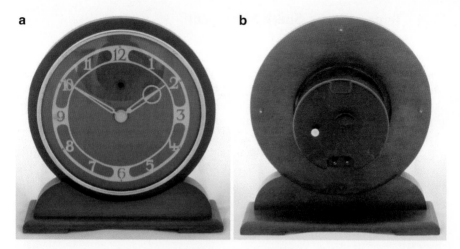

Fig. 7.61 Temco synchronous mantel clock, (**a**) Front, (**b**) Back

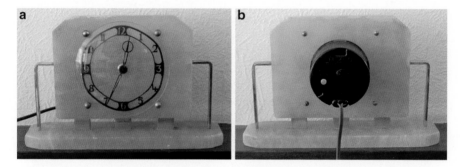

Fig. 7.62 Temco synchronous mantel clock, (**a**) Front, (**b**) Back

high × 19.5 cm wide. The dial is 12.5 cm diameter with a chrome bezel, a cream skeletonised chapter ring with Arabic numerals, and cream hands. The chapter ring is not quite circular and four of the numerals are cut outs. Brands on the back are Temco and Telephone Mfg. Co. Ltd. The clock is fitted with a Temco Mark V movement. To start the clock rotate the starter knob. There is a tell tale in an aperture at 12 o'clock. The rotating starter knob acts as a tell tale.

Temco Synchronous Mantel Clock See Fig. 7.62. For more information see Pook (2014). The clock is marked Made in England. It is for a 200–250 V supply. The style is Art Deco. The marble and chrome case is 20.5 cm high × 32 cm wide. The dial is 12.5 cm diameter with a black skeletonised chapter ring with Arabic numerals, and black hands. The chapter ring is not quite circular and four of the numerals are cut outs. Brands on the back are Temco and Telephone Mfg. Co. Ltd. The clock is fitted with a Temco Mark V movement. To start the clock rotate the starter knob. The rotating starter knob acts as a tell tale.

7.28   Temco Clocks, Smith Movement

Temco Synchronous Mantel Clock   See Fig. 7.63. The clock is marked Made in England. It is for a 200–250 V supply. The style is Chinoiserie. The lacquered wood case is 19 cm high × 26.5 cm wide. The dial is 12.5 cm diameter with a white skeletonised chapter ring with Arabic numerals, and white hands. The chapter ring is not quite circular and four of the numerals are cut outs. Brands on the back are SEC and Smiths English Clocks. The clock is fitted with a Smith Bijou movement. To start the clock press the starter knob and release. There is no tell tale. British Patent information on the clock is '366710, 369438, 374713, 384441, 484222'. The brand Temco does not appear on this clock. The main evidence for regarding this clock as a Temco fitted with a Smith movement is that the unusual skeletonised chapter ring does appear on other types of Temco clocks (Figs. 7.60, 7.61, and 7.62) but does not appear on Smith clocks.

Temco Chiming Mantel Clock   See Fig. 7.64. The clock is marked Made in England. It is for a 200–250 V supply. The wood veneer case is 21.5 cm high ×

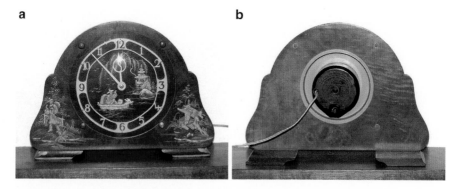

Fig. 7.63   Temco synchronous mantel clock, (**a**) Front, (**b**) Back

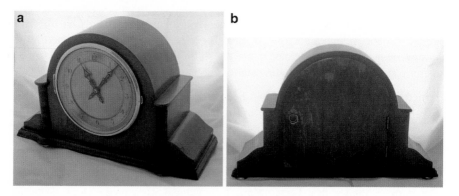

Fig. 7.64   Temco chiming mantel clock, (**a**) Front, (**b**) Back

35.5 cm wide. The dial is 12.5 cm diameter with a chrome bezel, and black Arabic numerals and hands. The brand on the back, with the door open, is Temco. The clock is fitted with a Smith Model 259 chiming movement, branded Temco. To start the clock push the starter lever and release. There is no tell tale. British patent information on the clock is '366710, 387108'.

7.29   Conclusions

1. The sample of synchronous mantel clocks illustrated cannot be regarded as representative of clocks that were produced, or have survived. Nevertheless, the statistics provide a rough indication for clocks that have survived.
2. The large number of Smith clocks (28 %) and Temco clocks (26 %) is noticeable.
3. A number of clocks (23 %) have Art Deco cases, but there are few clocks with other identifiable artistic styles.
4. Relatively few clocks (14 %) are self starting. None of them has an outage indicator.
5. A significant number of clocks (30 %) do not have any form of tell tale to make it easy for a user to determine whether or not a clock is running.
6. None of the clocks show the date, or the day of the week.

References

Anonymous (1940) Temco synchronous electric clocks. Horological J 82(984):14
Anonymous (1948) Review of the 1948 British industries fair. Horological section. Horological J 90(1076):280–283
Bird C (2003) Metamec. The clockmaker. Dereham. Antiquarian Horological Society, Ticehurst
British Patent 366710 (1932) Improvements relating to electric motors
British Patent 369438 (1932) Improvements relating to electric clocks
British Patent 374713 (1932) Improvements relating to electric clocks
British Patent 377696 (1932) Improvements in alternating current motors
British Patent 379383 (1932) Improvements in and relating to synchronous electric motors
British Patent 384441 (1932) Improvements relating to electric clocks
British Patent 384681 (1932) Improvements in and relating to synchronous electric motors
British Patent 384880 (1932) Improvements in electric motors
British Patent 387108 (1933) Electric clocks
British Patent 412336 (1934) Improvements relating to synchronous electric motors
British Patent 444688 (1936) Improvements in and relating to synchronous-motor electric clocks
British Patent 484222 (1938) Improvements relating to small synchronous electric motors
British Patent 744204 (1956) Improvements in or relating to the control of the rotational direction of synchronous electric motors
Lines MA (2012) Ferranti synchronous electric clocks, Paperback edition with corrections. Zazzo Media, Milton Keynes
Miles RHA (2011) Synchronome. Masters of electrical time keeping. The Antiquarian Horological Society, Ticehurst
Pook LP (2014) Temco Art Deco domestic synchronous clocks. Watch Clock Bull 56/1(407): 47–58
Smith B (2008) Smiths domestic clocks, 2nd edn. Pierhead Publications Limited, Herne Bay

Chapter 8
Gallery of Synchronous Bedside Clocks

Abstract A bedside clock is a clock intended for display on a bedside table, but which can also be used as a desk clock. A bedside clock has a dial that can be read from a short distance. The main purpose of this gallery is to illustrate the range of British synchronous bedside clocks that were available. A clock was assigned as a bedside clock primarily on the basis of the legibility of its dial. The dial on most of the clocks in this gallery is smaller than 9 cm. There are back and front views of each clock, together with a brief description. Clocks are mostly illustrated as found, and some are in poor condition. An alarm clock and a time switch are included. The sample of 43 synchronous bedside clocks, plus an alarm clock and a time switch illustrated cannot be regarded as representative of those that were produced, or have survived. Nevertheless, the statistics provide a rough indication for those that have survived. The large number of Smith clocks, an alarm clock and a time switch (31 %) and Temco clocks (29 %) is noticeable. A number of clocks etc. (31 %) have Art Deco cases, but there are few clocks with other identifiable artistic styles. Relatively few clocks etc. (13 %) are self starting. None of them has an outage indicator. Some clocks etc. (16 %) do not have any form of tell tale to make it easy for a user to determine whether or not a clock is running. None of the clocks etc. show the date, or the day of the week.

8.1 Introduction

A bedside clock is a clock intended for display on a bedside table, but which can also be used as a desk clock. A bedside clock has a dial that can be read from a short distance. An alarm clock is a bedside clock with a time switch and an audible alarm as complications. Some time switches can be used as bedside clocks. The main purpose of this gallery is to illustrate the range of British synchronous bedside clocks that were available. A clock was assigned as a bedside clock primarily on the basis of the legibility of its dial. The dial on most of the clocks in this gallery is smaller than 9 cm. There are back and front views of each clock, together with a brief description. Clocks are mostly illustrated as found, and some are in poor condition. An alarm clock and a time switch are included.

© Springer International Publishing Switzerland 2015   135
L.P. Pook, *British Domestic Synchronous Clocks 1930–1980*, History
of Mechanism and Machine Science 29, DOI 10.1007/978-3-319-14388-0_8

Fig. 8.1 Bowden & Sons synchronous alarm clock, (**a**) Front, (**b**) Back

8.2 Bowden & Sons Clock

Bowden & Sons Synchronous Alarm Clock See Fig. 8.1. This is a rebranded Smith *Callboy*, date range 1933–1950 (Smith 2008). The clock is marked Made in England. It is for a 200–250 V supply. The style is Bakelite. The Bakelite case is 12 cm high × 11 cm wide. The dial is 6.5 cm high × 6.5 cm wide with luminous Arabic numerals, and black hands with luminous tips The brand on the front is Bowden & Sons and the brand on the back is Smiths English Clocks Ltd. The clock is fitted with a Smith self starting Callboy movement. There is no tell tale. British Patent information on the clock is '366710, 369438, 374713, 402456, 419767, others pending'.

8.3 Elco Clock

Elco Synchronous Bedside Clock See Fig. 8.2. The clock is marked Made in Great Britain. It is for a 200–250 V supply. The oak case is 13.5 cm high × 19 cm wide. The dial is 8.5 cm square with cream Arabic numerals and chrome hands. The brand on the front is Elco. The brand on the back is Elco Clocks & Watches Ltd. The clock is fitted with an Elco movement. To start the clock press the starter knob and release. There is a tell tale in an aperture at 12 o'clock.

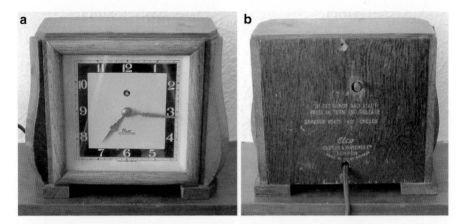

Fig. 8.2 Elco synchronous bedside clock, (**a**) Front, (**b**) Back

8.4 Ferranti Clocks

Model numbers and dates shown are based on information in Lines (2012). There are some inconsistencies. In particular, Ferranti sometimes used the same model number for different clocks produced at different times. It is not clear whether this was intentional or an oversight (Lines, personal communication 2012). There are often variations, for example in hand style, for clocks with the same model number produced in different years.

Ferranti Model No. 3 Synchronous Bedside Clock See Figs. 6.1a and 8.3. For more information see Lines (2012). The date range is 1932–1938. The clock is marked Made in England. It is for a 200–250 V supply. The style is Art Deco. The Bakelite case is 15.5 cm high × 15 cm wide. The dial is 8 cm diameter with black Arabic numerals and hands, and a brass seconds hand. The brand on the front is Ferranti.The clock is fitted with a Ferranti Early movement. To start the clock rotate the starter knob. The inconspicuous seconds hand and rotating starter knob act as tell tales.

Ferranti Model No. 5 Synchronous Bedside Clock See Fig. 8.4. For more information see Lines (2012). The date range is 1932–1938. The clock is marked Made in England. It is for a 200–250 V supply. The style is Bakelite. The Bakelite case is 15 cm high × 18 cm wide. The dial is 8.5 cm diameter with a chrome bezel, and black Arabic numerals and hands. The dial is the same as the 1936 version but the hands are the same as the 1938 version, except that there is no seconds hand. The brand on the front and on the back is Ferranti. The clock is fitted with a Ferranti Later movement. To start the clock rotate the starter knob. The rotating starter knob acts as a tell tale. British Patent information on the clock is 'Patents pending'.

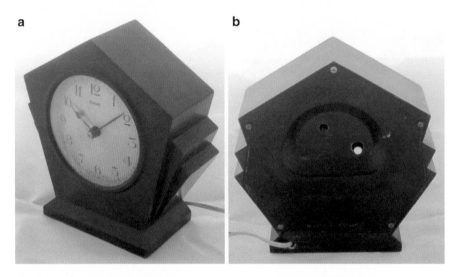

Fig. 8.3 Ferranti Model No. 3 synchronous bedside clock, (**a**) Front, (**b**) Back

Fig. 8.4 Ferranti Model No. 5 synchronous bedside clock, (**a**) Front, (**b**) Back

Ferranti Model No. 12 Synchronous Bedside Clock See Figs. 4.7b and 8.5. For more information see Lines (2012). The date range is 1934–1936. The same model number was also used for a different, later clock. The clock is marked Made in England. It is for a 200–250 V supply. The style is Geometric Octagonal. The hammered pewter case is 12.5 cm high × 12 cm wide. The dial is 7.5 cm diameter with a chrome bezel, black Arabic numerals and hands, and a white seconds hand. The brand on the front and on the back, with the cover removed, is Ferranti. The brand on the base is Talbot Pewter. The clock is fitted with a Ferranti Early

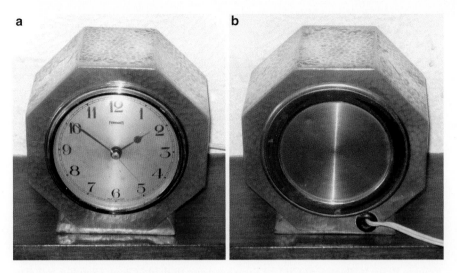

Fig. 8.5 Ferranti Model No. 12 synchronous bedside clock, (**a**) Front, (**b**) Back

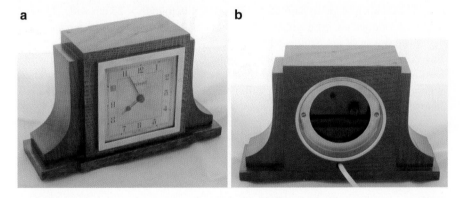

Fig. 8.6 Ferranti Model No. 132 synchronous bedside clock, (**a**) Front, (**b**) Back

movement. To start the clock rotate the starter knob. The inconspicuous seconds hand and rotating starter knob act as tell tales. British Patent information on the clock is 'Patents pending'.

Ferranti Model No. 132 Synchronous Bedside Clock See Fig. 8.6. For more information see Lines (2012). This is the 1936 version. The clock is marked Made in England. It is for a 200–250 V supply. The oak case is 14.5 cm high × 23.5 cm wide. The dial is 8.5 cm square with a chrome bezel, black Arabic numerals and hands, and a brass seconds hand. The brand on the front and on the back is Ferranti. The clock is fitted with the Ferranti Early movement. To start the clock rotate the starter knob. The inconspicuous seconds hand and rotating starter knob act as tell tales.

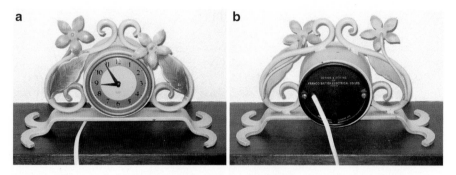

Fig. 8.7 Franco British Electrical Co Ltd synchronous bedside clock, (**a**) Front, (**b**) Back

8.5   Franco British Electrical Co Ltd Clock

Franco British Electrical Co Ltd Synchronous Bedside Clock   See Fig. 8.7. The clock is marked Made in England. It is for a 200–250 V supply. The style is Wrought Iron. The cream and green painted metal case is 14 cm high × 23 cm wide. The dial is 6.5 cm diameter with a gilt bezel, and black Arabic numerals and hands. The brands on the front are Smiths and SECTRIC. The brands on the back are Franco British Electrical Co. Ltd and Smith's English Clocks Ltd. The clock is fitted with a Smith Bijou movement. To start the clock press the starter knob and release. There is no tell tale. British Patent information on the clock is '366710, 369438, 374713, 384441, 484222'.

8.6   GEC Clock

GEC Synchronous Bedside Clock   See Fig. 8.8. The clock is marked Made in England. It is for a 200–250 V supply. The style is Bakelite. The brand on the front is GEC, and the brands on the back are SEC and General Electric Co Ltd. The Bakelite case is 16 cm high × 13.5 cm wide. The case is the same as the Smith *Daventry* and Smith *Tenby* clocks (Smith (2008). The date for both of these is 1933. The dial is 8.5 cm diameter with a brass bezel, black Arabic numerals and hands and a seconds indicator in an aperture at 6 o'clock. The dial is the same as the Smith *Tenby* and the hands the same as the Smith *Daventry*. The clock is fitted with a Smith Type 1 movement. To start the clock press the starter knob and release. The seconds indicator acts as a tell tale. British patent information on the clock is 'Patents pending'.

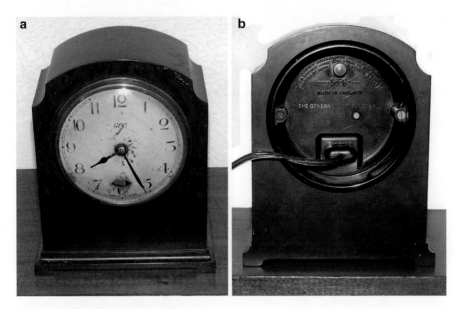

Fig. 8.8 GEC synchronous bedside clock, (**a**) Front, (**b**) Back

8.7 Genalex Clock

Genalex Synchronous Bedside Clock See Fig. 8.9. The clock is marked Made in England. It is for a 200–250 V supply. The inlaid veneer case with brass feet is 14.5 cm high × 18 cm wide. The dial is 8.5 cm diameter with a brass bezel, and black Arabic numerals and hands. The brand on the front and on the back is Genalex. The clock is fitted with a Smith BM7 movement. To start the clock press the starter knob and release. There is a tell tale in an aperture at 12 o'clock. British Patent information on the clock is '366710, 369438, 374713, others pending'.

8.8 Goblin Clock

Goblin Model 394 Synchronous Bedside Clock See Figs. 4.7a and 8.10. The clock is marked British Made. It is for a 200–250 V supply. The style is Geometric Hexagonal. The onyx case is 11 cm high × 12 cm wide. The dial is 8.5 cm diameter with a chrome bezel, black Arabic numerals and hands, and a red seconds hand. The brand on the front and on the back is Goblin. The clock is fitted with a Goblin self starting movement. The seconds hand acts as a tell tale. British Patent information on the clock is '366710, 571849'.

Fig. 8.9   Genalex synchronous bedside clock, (**a**), Front, (**b**) Back

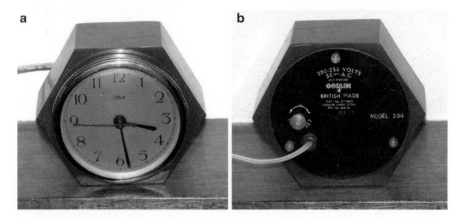

Fig. 8.10   Goblin Model 394 synchronous bedside clock, (**a**) Front, (**b**) Back

8.9   Kelnore Clock

Kelnore Synchronous Bedside Clock   See Fig. 8.11. The clock is marked Made in England. It is for a 200–250 V supply. The style is Bakelite. The Bakelite case is 15 cm high × 19 cm wide. The dial is 9 cm diameter with a chrome bezel, black Arabic numerals and hands, and a brass seconds hand. The brand on the front is Kelnore. The brand on the back is J. H. Jerrim & Co Ltd. The clock is fitted with a Kelnore movement. To start the clock rotate the starter knob. The seconds hand and rotating starter knob act as tell tales.

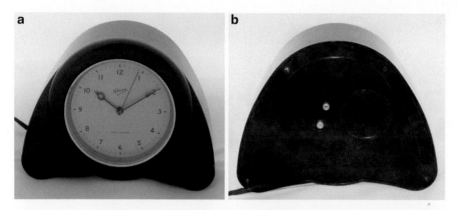

Fig. 8.11 Kelnore synchronous bedside clock, (**a**) Front, (**b**) Back

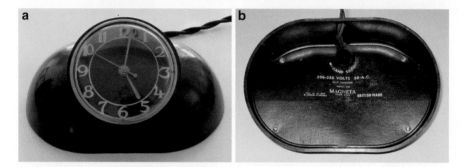

Fig. 8.12 Magneta Model 650 synchronous bedside clock, (**a**) Top, (**b**) Bottom

8.10 Magneta Clock

Magneta Model 650 Synchronous Bedside Clock See Fig. 8.12. The clock is British made. It is for a 200–250 V supply. The style is Bakelite. The Bakelite case is 9 cm high × 18 cm wide. The dial is 7.5 cm diameter with a brass bezel, and brass Arabic numerals and hands. The clock is fitted with a Magneta self starting Later movement. The seconds hand acts as tell tale. British Patent information on the clock is '571849, others pending'.

8.11 Metamec Clocks

Metamec Model 7551 Synchronous Bedside Clock See Fig. 8.13. For more information see Bird (2003). The clock is marked Made in England. It is for a 200–250 V supply. The style is Pictorial. The cream plastic case is 14 × 14 cm.

Fig. 8.13 Metamec Model 7551 synchronous bedside clock, (**a**) Front, (**b**) Back

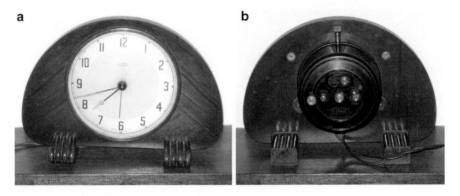

Fig. 8.14 Metamec synchronous alarm clock, (**a**) Front, (**b**) Back

The dial is 10 cm diameter with white Arabic numerals, and black hands. The brand on the front and on the back is Metamec. The clock is fitted with a Metamec Type 1 movement. To start the clock turn the starter knob past the click. The rotating windmill sails and optional artificial click acts as tell tales.

Metamec Synchronous Alarm Clock See Fig. 8.14. This is the alarm version of Model 746W (Bird 2003). The clock is marked Made in England. It is for a 200–250 V supply. The wood veneer case is 14 cm high × 20.5 cm wide. The dial is 11.5 cm diameter with a brass bezel, black Arabic numerals, luminous hour dots, black hands with luminous tips and a red alarm hand. The dial is unusually large for an alarm clock. The brand on the front and on the back is Metamec. The clock is fitted with an alarm version of the Metamec Type 1 movement. To start the clock press the starter knob and release. The optional artificial tick acts as a tell tale.

8.12 Smith Clocks, Type 1 Movement

Smith *Arundel* Synchronous Bedside Clock See Fig. 8.15 The date range is 1932–1933 (Smith 2008). The clock is marked Made in England. It is for a 200–250 V supply. The style is Art Deco. The mahogany case is 16 cm high × 19.5 cm wide. The dial is 8.5 cm diameter with a chrome bezel, black Arabic numerals and hands. There is a seconds indicator in an aperture at 6 o'clock. The brands on the front are Smith and SECTRIC. The brands on the back are SEC, Smith's English Clocks Ltd, and Synchronous Electric Clocks Ltd. A paper label on the base of the clock reads 'GEO. COULBECK, Wholesale Fish Merchant, FISH DOCKS, GRIMSBY'. The clock is fitted with a Smith Type 1 movement. To start the clock press the starter knob and release. The seconds indicator acts as a tell tale. British Patent information on the clock is 'Patents pending'.

Smith *Delta* Synchronous Bedside Clock See Figs. 4.1b and 8.16. The date is 1933 (Smith 2008). The clock is marked Made in England. It is for a 200–250 V supply. The style is Art Deco. The oak case is 14.5 cm high × 18 cm wide. The dial is 7.5 cm diameter, with a brass bezel, and black Arabic numerals and hands. There is a seconds indicator in an aperture at 6 o'clock. The brand on the front is Smith. The brands on the back are SEC, Smith's English Clocks Ltd, and Synchronous Electric Clocks Ltd. The clock is fitted with a Smith Type 1 movement. To start the clock press the starter knob and release. The seconds indicator acts as a tell tale. British Patent information on the clock is '366710, 369438, 374713, others pending'.

Smith *Norwich* Synchronous Bedside Clock See Figs. 4.9 and 8.17 The date range is 1932–1933 (Smith 2008). The clock is marked Made in England. It is for a 200–250 V supply. The style is Napoleon's Hat. The wood veneer case is 14 cm high × 31 cm wide. The dial is 8.5 cm diameter with a chrome bezel, and black Arabic

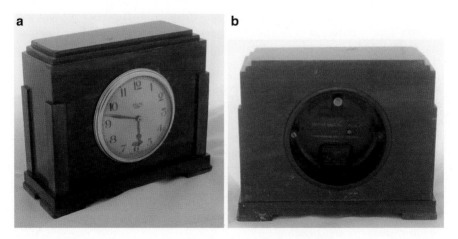

Fig. 8.15 Smith *Arundel* synchronous bedside clock, (**a**) Front. (**b**) Back

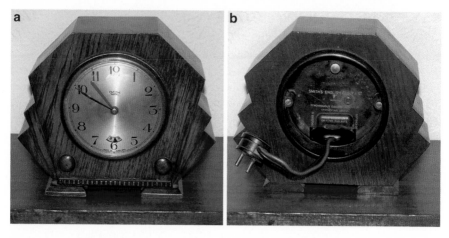

Fig. 8.16 Smith *Delta* synchronous bedside clock, (**a**) Front, (**b**) Back

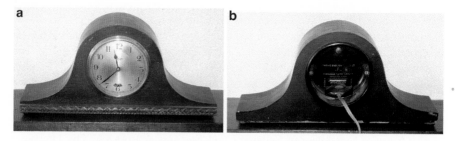

Fig. 8.17 Smith *Norwich* synchronous bedside clock, (**a**) Front, (**b**) Back

numerals and hands. There is a seconds indicator in an aperture at 6 o'clock. The brand on the front is Smith. The brands on the back are Smith, SECTRIC, Smith's English Clocks Ltd, and Synchronous Electric Clocks Ltd. The clock is fitted with a Smith Type 1 movement. To start the clock press the starter knob and release. The seconds indicator acts as a tell tale. British Patent information on the clock is '366710, 369438, 374713, others pending'.

Smith Synchronous Bedside Clock See Fig. 8.18. The case style is similar to the Bakelite case of the Smith *Rugby* clock, date range 1932–1933 (Smith 2008). The clock is marked Made in England. It is for a 200–250 V supply. The inlaid wood veneer and brass case is 20.5 cm high × 15 cm wide. The dial is 9 cm diameter with a brass bezel, black Arabic numerals and hands. There is a seconds indicator in an aperture at 6 o'clock. The brand on the front is Smith. The brands on the back are SECTRIC, Smith's English Clocks Ltd, and Synchronous Electric Clocks Ltd. The clock is fitted with a Smith Type 1 movement. To start the clock press the starter knob and release. The seconds indicator acts as a tell tale. British Patent information on the clock is 'Patents pending'.

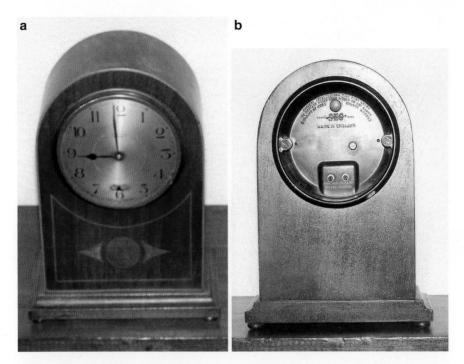

Fig. 8.18 Smith synchronous bedside clock, (**a**) Front, (**b**) Back

8.13   Smith Clocks, BM7 Movement

Smith *Albury* Synchronous Bedside Clock   See Fig. 8.19. The date range is 1935–1939 (Smith 2008). The clock is marked Made in England. It is for a 200–250 V supply. The oak case is 14 cm high × 24 cm wide. The dial is 8.5 cm diameter with a chrome bezel, and black Arabic numerals and hands. The brand on the front is Smith. Brands on the back are SEC, Smith's English Clocks Ltd, and Synchronous Electric Clocks Ltd. The clock is fitted with a Smith BM7 movement. To start the clock press the starter knob and release. There is a tell tale in an aperture at 12 o'clock. British Patent information on the clock is '366710, 369438, 374713, others pending'.

Smith *Capri* Synchronous Bedside Clock   See Fig. 8.20. The date is 1937 (Smith 2008). The clock is marked Made in England. It is for a 200–250 V supply. The style is Art Deco. The chrome and Bakelite case is 13.5 cm high × 12 cm wide. The dial is 7.5 cm diameter with a chrome bezel, and black Arabic numerals and hands. The brand on the front is Smith. Brands on the back are SEC, Smith's English Clocks Ltd, and Synchronous Electric Clocks Ltd. The clock is fitted with a Smith BM7

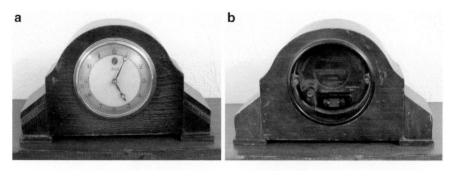

Fig. 8.19 Smith *Albury* synchronous bedside clock, (**a**) Front, (**b**) Back

Fig. 8.20 Smith *Capri* synchronous bedside clock, (**a**) Front, (**b**) Back

movement. To start the clock press the starter knob and release. There is a tell tale in an aperture at 12 o'clock. British Patent information on the clock is '366710, 369438, 374713, others pending'.

Smith *Creetown Variant* Synchronous Bedside Clock See Fig. 8.21. Except for the base this is the same as Smith *Creetown*, date range 1935–1936 (Smith 2008). The clock is marked Made in England. It is for a 200–250 V supply. The style is Art Deco. The chrome and ebonite case is 16.5 cm high × 14 cm wide. The dial is 9 cm diameter with black Roman numerals, and chrome hands. The brands on the front are Smith and SECTRIC. The brands on the back are SEC, Smiths English Clocks Ltd, and Synchronous Electric Clocks Ltd. The clock is fitted with a Smith BM7 movement. To start the clock press the starter knob and release. There is a tell tale in an aperture at 12 o'clock. British Patent information on the clock is '366710, 369438, 374713, others pending'.

Fig. 8.21 Smith *Creetown* synchronous bedside clock, (**a**) Front, (**b**) Back

Fig. 8.22 Smith *Darwin* synchronous bedside clock, (**a**) Front, (**b**) Back

Smith *Darwin* Synchronous Bedside Clock See Fig. 8.22. The date range is 1935–1939 (Smith 2008). The clock is marked Made in England. It is for a 200–250 V supply. The oak case is 14 cm high × 20.5 cm wide. The dial is 7.5 cm square with a chrome bezel, and black Arabic numerals and hands. The brands on the front are Smith and SECTRIC. The brands on the back are SEC, Smiths English Clocks Ltd, and Synchronous Electric Clocks Ltd. The clock is fitted with a Smith BM7 movement. To start the clock press the starter knob and release. There is a tell tale in an aperture at 12 o'clock. British Patent information on the clock is '366710, 369438, 374713, others pending'.

8.14   Smith Clocks, Bijou Movement

Smith *Cleveforth* Synchronous Bedside Clock   See Fig. 8.23. The date range is 1957–1958 (Smith 2008). The clock is marked Made in England. It is for a 200–250 V supply. The brass case is 11 cm high × 19 cm wide. The dial is 8.5 cm high × 11.5 cm wide, with brass batons and hands. The brands on the front are Smiths and SECTRIC. The brands on the back are Smiths English Clocks and SEC. The clock is fitted with a Smith Bijou movement. To start the clock press the starter knob and release. There is no tell tale. British Patent Information on the clock is '366710, 369438, 374713, 384441, 484222'.

Smith *Cromead* Synchronous Bedside Clock   See Fig. 8.24. The date is 1950 (Smith 2008). The clock is marked Made in England. It is for a 200–250 V supply. The style is Bakelite. The Bakelite case is 11 cm high × 19 cm wide. The dial is 9 cm diameter with a brassed bezel, and black Roman numerals and hands. The brands

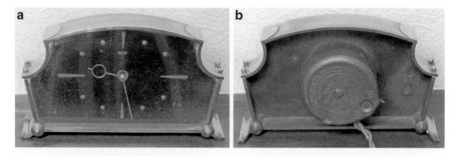

Fig. 8.23   Smith *Cleveforth* synchronous bedside clock, (**a**) Front, (**b**) Back

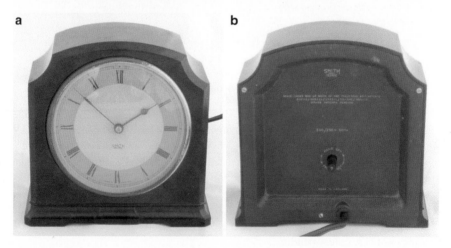

Fig. 8.24   Smith *Cromead* synchronous bedside clock, (**a**) Front, (**b**) Back

a   b

Fig. 8.25 Smith *Heston* synchronous bedside clock, (**a**) Front, (**b**) Back

on the front are Smith and SECTRIC. The brands on the back are Smiths and
SECTRIC. The clock is fitted with a Smith Bijou movement. To start the clock
press the starter knob and release. There is no tell tale. British Patent Information
on the clock is '366710, 369438, 374713, 384441, 484222, others pending'.

Smith *Heston* Synchronous Bedside Clock See Fig. 8.25. The date range is
1938–1939 (Smith (2008). The clock is marked Made in England. It is for a
200–250 V supply. The wood veneer case is 14 cm high × 14.5 cm wide. The dial is
9 cm diameter with white Arabic numerals and brass hands. The clock is fitted with
a Smith Bijou movement. To start the clock press the starter knob and release. There
is no tell tale. British Patent Information on the clock is '366710, 369438, 374713,
others pending'.

Smith *Nell Gwynne* Synchronous Bedside Clock See Figs. 6.1b and 8.26. The date
range is 1950–1978 (Smith 2008). The clock is marked Made in England. It is for
a 200–250 V supply. The clock is a reproduction of a traditional style. The brassed
steel case is 16.5 cm high × 7.5 cm wide. The unglazed dial is 7.5 cm diameter
with black Roman numerals and hands. The brand on the back, with the door open,
is Smiths English Clocks Ltd. The clock is fitted with a Smith self starting Bijou
movement. With the door open the visible rotor acts as a tell tale (Fig. 11.32d).

Smith *Prestwick* Synchronous Time Switch See Fig. 8.27. The date is 1950 (Smith
2008). The maximum output load is 300 W. The time switch is also branded EKCO,
a radio manufacturer. The time switch turns on and off once per day. The intention
was that it would be used to turn a radio on and off. The time switch is marked Made
in England. It is for a 200–250 V supply. The wood veneer case is 12.5 cm high ×
16 cm wide. The dial is 8 cm diameter with black Arabic numerals and hands. There

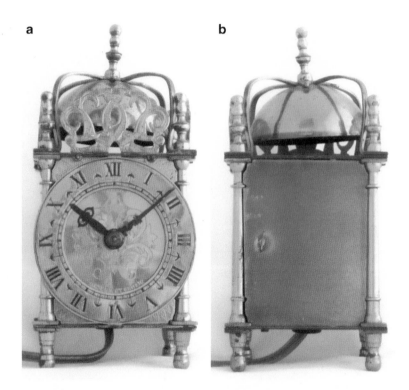

Fig. 8.26 Smith *Nell Gwynne* synchronous bedside clock, (**a**) Front, (**b**) Back

Fig. 8.27 Smith *Prestwick* synchronous time switch, (**a**) Front, (**b**) Rear

is an aperture for an AM/PM indicator at 12 o'clock, and apertures for time switch settings at 3 o'clock and 9 o'clock. Brands on the front are EKCO and SECTRIC. Brands on the back are EKCO, E. K. Cole Ltd, Smith, and SECTRIC. The time switch is fitted with a Smith self starting Bijou movement. There is no tell tale.

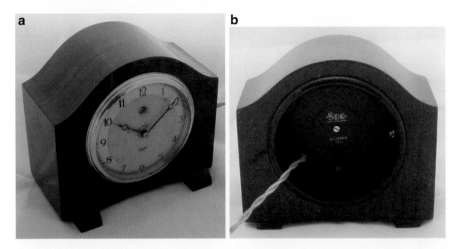

Fig. 8.28 Smith synchronous bedside clock, (**a**) Front, (**b**) Back

Smith Synchronous Bedside Clock See Fig. 8.28. The clock is marked Made in England. It is for a 200–250 V supply. The walnut veneer case is 12.5 cm high × 14.5 cm wide. The dial is 7.5 cm diameter with a chrome bezel, and black Arabic numerals and hands. The brands on the front are Smith and SECTRIC. The brand on the back is SEC. The clock is fitted with a Smith Bijou movement. To start the clock press the starter knob and release. There is a tell tale in an aperture at 12 o'clock. British patent information on the clock is '369438, 374713, others pending'.

8.15 Smith Clock, QGEM Movement

Smith *Phillipe* Synchronous Bedside Clock See Figs. 4.3 and 8.29. The date is 1978 (Smith 2008). The clock is marked Made in England. It is for a 200–250 V supply. The style is Carriage Clock The brass case is 17 cm high × 9.5 cm wide. The dial is 7 cm diameter with black Roman numerals and hands. The brand on the front is Smiths. The brands on the back, with the door open, are Smiths Clocks and Watches Ltd, and S Smith and Sons (England) Ltd. The clock is fitted with a Smith self starting QGEM movement. There is no tell tale. British Patent information on the clock is '744204, 806383'.

8.16 Sterling Clocks

Sterling Synchronous Bedside Clock See Fig. 8.30. The clock is marked Made in England. It is for a 200–250 V supply. The style is Art Deco. The oak case is 13.5 cm high × 18 cm wide. The dial is 9 cm square with a chrome bezel, and black Arabic

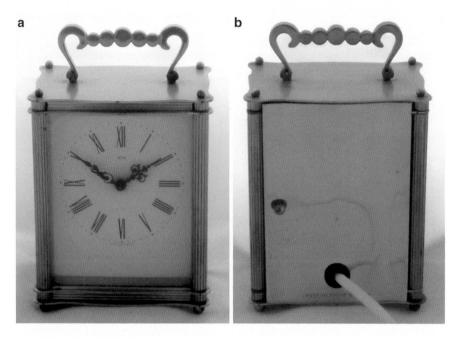

Fig. 8.29 Smith *Phillipe* synchronous bedside clock, (**a**) Front, (**b**) Back

Fig. 8.30 Sterling synchronous bedside clock, (**a**) Front, (**b**) Back

numerals and hands. The brand on the front is Sterling. The brand on the back is Sterling – Croydon Clocks Ltd. The clock is fitted with a Sterling movement. To start the clock twist the starter knob and release. The seconds hand acts as a tell tale.

Sterling Synchronous Bedside Clock   See Fig. 8.31. The clock is marked Made in England. It is for a 220 V supply. The style is Art Deco. The oak case is 14.5 cm high × 18 cm wide. The dial is 8.5 cm square with a chrome bezel, black Arabic

Fig. 8.31 Sterling synchronous bedside clock, (**a**) Front, (**b**) Back

numerals and hands, and a chrome seconds hand. The brand on the front is Sterling. The clock is fitted with a Sterling movement. To start the clock twist the starter knob and release. The seconds hand acts as a tell tale.

8.17 Synchronomains Clock

Synchronomains Synchronous Bedside Clock See Figs. 4.2a and 8.32. The clock was advertised in December 1931 (Miles 2011). The case is the same as that of the Smith *Daventry*, date range 1932–1933 (Smith 2008), but the dial is different. The clock is marked Made in England. It is for a 200–250 V supply. The style is Bakelite. The Bakelite case is 16 cm high × 13.5 cm wide. The dial is 7.5 cm diameter with a chrome bezel, and black Arabic numerals and hands. The brand on the front and on the back is Synchronomains. The clock is fitted with an Empire movement, serial number 12274. To start the clock push the starter lever and release. There is a tell tale in an aperture at 6 o'clock. British Patent information on the clock is '386756, patent pending'.

8.18 Temco Clock, Mark II Movement

Temco Synchronous Bedside Clock See Fig. 8.33. The clock is marked Made in England. It is for a 200–240 V supply. The inlaid walnut veneer case is 15 cm high × 18.5 cm wide. The dial is 8.5 cm diameter with a brass bezel, and black Arabic numerals and hands. There is an aperture for a seconds indicator at 5 o'clock.

Fig. 8.32 Synchronomains synchronous bedside clock, (**a**) Front, (**b**) Back

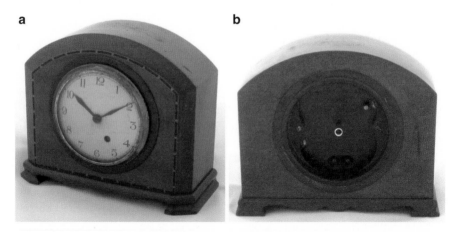

Fig. 8.33 Temco synchronous bedside clock, (**a**) Front, (**b**) Back

Brands on the back are Temco and Telephone Mfg. Co. Ltd. '224' is stamped on the base. This is probably the model number. The clock is fitted with a Temco Mark II movement, serial number 15708. To start the clock press the starter knob and release. The seconds indicator acts as a tell tale. British Patent information on the clock is 'Patents pending'.

8.19 Temco Clocks, Mark IV Movement

Temco Synchronous Bedside Clock See Fig. 8.34. The clock is marked Made in England. It is for a 200–250 V supply. The style is Art Deco. The walnut case is 16.5 cm high × 29 cm wide. The dial is 7.5 cm square with a chrome bezel, and black Arabic numerals and hands. Brands on the back are Temco and Telephone Mfg. Co. Ltd. The clock is fitted with a Temco Mark IV movement, serial number 10376. To start the clock rotate the starter knob. The rotating starter knob acts as a tell tale.

Temco Synchronous Bedside Clock See Fig. 8.35. The clock is marked Made in England. It is for a 200–250 V supply. The wood veneer case is 12.5 cm high × 25.5 cm wide. The dial is 8.5 cm diameter with a chrome bezel, and black Arabic numerals and hands. Brands on the back are Temco and Telephone Mfg. Co. Ltd. The clock is fitted with a Temco Mark IV movement, serial number 32704. To start the clock rotate the starter knob. There is a tell tale in an aperture at 12 o'clock. The rotating starter knob acts as a tell tale.

Temco Synchronous Bedside Clock See Fig. 8.36. The clock is marked Made in England. It is for a 200–250 V supply. The oak veneer case is 16.5 cm high × 21.5 cm high. The dial is 7.5 cm diameter with a chrome bezel, and black Arabic

Fig. 8.34 Temco synchronous bedside clock, (**a**) Front, (**b**) Back

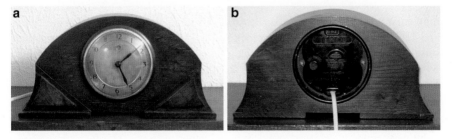

Fig. 8.35 Temco synchronous bedside clock, (**a**) Front, (**b**) Back

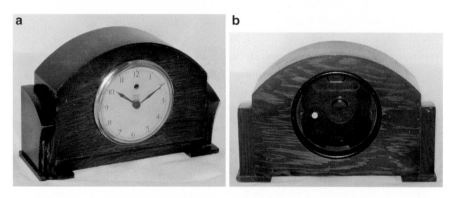

Fig. 8.36 Temco synchronous bedside clock, (**a**) Front, (**b**) Back

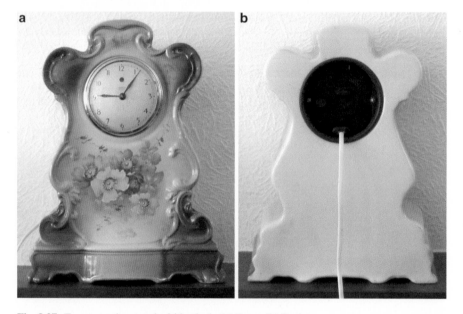

Fig. 8.37 Temco synchronous bedside clock, (**a**) Front, (**b**) Back

numerals and hands. The brand on the front is Temco. Brands on the back are Temco and Telephone Mfg. Co. Ltd. The clock is fitted with a Temco Mark IV movement, serial number 37922. To start the clock rotate the starter knob. There is a tell tale in an aperture at 12 o'clock. The rotating starter knob acts as a tell tale.

Temco Synchronous Bedside Clock See Fig. 8.37. The clock is marked Made in England. It is for a 200–250 V supply. The porcelain case is 30.5 cm high × 23.5 cm wide. The dial is 8.5 cm diameter with a chrome bezel, and black Arabic numerals and hands. The brand on the front is Temco. Brands on the back are Temco and

Telephone Mfg. Co. Ltd. The clock is fitted with a Temco Mark IV movement. To start the clock rotate the starter knob. There is a tell tale in an aperture at 12 o'clock. The rotating starter knob acts as a tell tale.

8.20   Temco Clocks, Mark V Movement

Temco Synchronous Bedside Clock See Fig. 8.38. The clock is marked Made in England. It is for a 200–250 V supply. The style is Art Deco. The translucent amber plastic case is 11.5 cm high × 11.5 cm wide. The dial is 7.5 cm diameter with a brassed bezel, and black Arabic numerals and hands. Brands on the back are Temco and Telephone Mfg. Co. Ltd. The clock is fitted with a Temco Mark V movement. The movement is marked '21 JUL 1938', presumably the date of manufacture. To start the clock rotate the starter knob. There is a tell tale in an aperture at 12 o'clock. The rotating starter knob acts as a tell tale.

Temco Synchronous Bedside Clock See Fig. 8.39. The clock is marked Made in England. It is for a 200–250 V supply. The style is Art Deco. The dial is attached to a Bakelite stand by swivels. With the dial upright the case is 13 cm high × 16.5 cm wide. The dial is 7.5 cm square with a chrome bezel, and black Arabic numerals and hands. The brand on the front is Temco. Brands on the back are Temco and Telephone Mfg. Co. Ltd. The clock is fitted with a Temco Mark V movement. To start the clock rotate the starter knob. There is a tell tale in an aperture at 12 o'clock. The rotating starter knob acts as a tell tale.

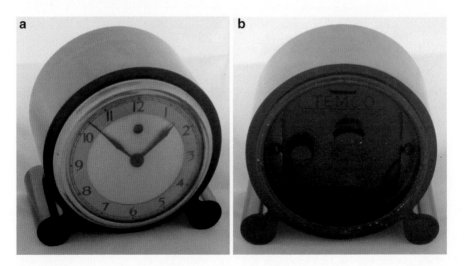

Fig. 8.38 Temco synchronous bedside clock, (**a**) Front, (**b**) Back

Fig. 8.39 Temco synchronous bedside clock, (**a**) Front, (**b**) Back

Fig. 8.40 Temco synchronous bedside clock, (**a**) Front, (**b**) Back

Temco Synchronous Bedside Clock See Fig. 8.40. The clock is marked Made in England. It is for a 200–250 V supply. The dial is attached to a stand by swivels. With the dial upright the cream plastic and chrome case is 12 cm high × 16.5 cm wide. The dial is 7.5 cm diameter with a chrome bezel, and black Arabic numerals and hands. The brand on the front is Temco. Brands on the back are Temco and Telephone Mfg. Co. Ltd. The clock is fitted with a Temco Mark V movement. To start the clock rotate the starter knob. There is a tell tale in an aperture at 12 o'clock. The rotating starter knob acts as a tell tale.

Temco Synchronous Bedside Clock See Fig. 8.41. The clock is marked Made in England. It is for a 200–250 V supply. The style is Art Deco. The chrome and black plastic case is 13.5 cm high × 15 cm wide. The dial is 7.5 cm square with a chrome bezel, and black Arabic numerals and hands. The brand on the front is Temco. Brands on the back are Temco and Telephone Mfg. Co. Ltd. The clock is

Fig. 8.41 Temco synchronous bedside clock, (**a**) Front, (**b**) Back

Fig. 8.42 Temco synchronous bedside clock, (**a**) Front, (**b**) Back

fitted with a Temco Mark V movement. To start the clock rotate the starter knob. There is a tell tale in an aperture at 12 o'clock. The rotating starter knob acts as a tell tale.

Temco Synchronous Bedside Clock See Fig. 8.42. The clock is marked Made in England. It is for a 200–250 V supply. The oak veneer case is 14.5 cm high × 19.5 cm wide. The dial is 7.5 cm diameter with black Arabic numerals and hands. The brand on the front is Temco. Brands on the back are Temco and The Telephone Mfg. Co. Ltd. The clock is fitted with a Temco Mark V movement. To start the clock rotate the starter knob. There is a tell tale in an aperture at 12 o'clock. The rotating starter knob acts as a tell tale.

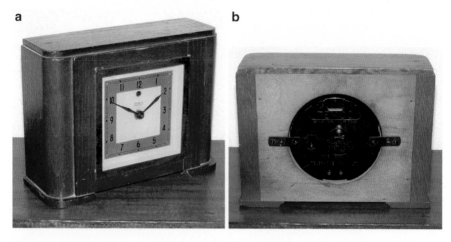

Fig. 8.43 Temco synchronous bedside clock, (**a**) Front, (**b**) Back

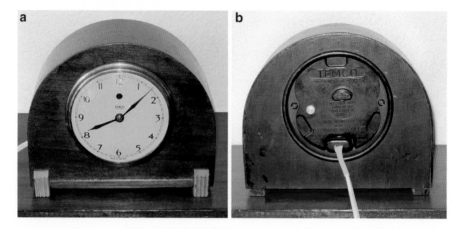

Fig. 8.44 Temco synchronous bedside clock, (**a**) Front, (**b**) Back

Temco Synchronous Bedside Clock See Fig. 8.43. The clock is marked Made in England. It is for a 200–250 V supply. The partly veneered wood case is 14.5 cm high × 19.5 cm wide. The dial is 7.5 cm square with a chrome bezel and black Arabic numerals and hands. The brand on the front is Temco. Brands on the back are Temco and The Telephone Mfg. Co. Ltd. The clock is fitted with a Temco Mark V movement. To start the clock rotate the starter knob. There is a tell tale in an aperture at 12 o'clock. The rotating starter knob acts as a tell tale.

Temco Synchronous Bedside Clock See Figs. 6.1c and 8.44. For more information see Pook (2014). The clock is marked Made in England. It is for a 200–250 V supply. The style is Art Deco. The wood veneer case is 16.5 cm high × 16.5 cm wide. The dial is 8.5 cm diameter with a chrome bezel, and black Arabic numerals

Fig. 8.45 Temco synchronous bedside clock, (**a**) Front, (**b**) Back (7 May 10 D, 7 May 10 E)

and hands. The brand on the front is Temco. Brands on the back are Temco and The Telephone Mfg. Co. Ltd. The clock is fitted with a Temco Mark V movement. To start the clock rotate the starter knob. There is a tell tale in an aperture at 12 o'clock. The rotating starter knob acts as a tell tale.

Temco Synchronous Bedside Clock See Fig. 8.45. The clock is marked Made in England. It is for a 200–250 V supply. The style is Art Deco. The oak case is 15 cm high × 20.5 cm wide. The dial is 8.5 cm square with a chrome bezel, and black Arabic numerals and hands. The brand on the front is Temco. Brands on the back are Temco and The Telephone Mfg. Co. Ltd. The clock is fitted with a Temco Mark V movement. To start the clock rotate the starter knob. There is a tell tale in an aperture at 12 o'clock. The rotating starter knob acts as a tell tale.

8.21 Conclusions

1. The sample of synchronous bedside clocks, an alarm clock and a time switch illustrated cannot be regarded as representative of those that were produced, or have survived. Nevertheless, the statistics provide a rough indication for those that have survived.
2. The large number of Smith clocks, an alarm clock and a time switch (31 %) and Temco clocks (29 %) is noticeable.
3. A number of clocks etc. (31 %) have Art Deco cases, but there are few clocks with other identifiable artistic styles.
4. Relatively few clocks etc. (13 %) are self starting. None of them has an outage indicator.
5. Some clocks etc. (16 %) do not have any form of tell tale to make it easy for a user to determine whether or not a clock is running.
6. None of the clocks etc. show the date, or the day of the week.

References

Bird C (2003) Metamec. The clockmaker. Dereham. Antiquarian Horological Society, Ticehurst

British Patent 366710 (1932) Improvements relating to electric motors

British Patent 369438 (1932) Improvements relating to electric clocks

British Patent 374713 (1932) Improvements relating to electric clocks

British Patent 384441 (1932) Improvements relating to electric clocks

British Patent 386756 (1933) Synchronous electric motors and means for starting them

British Patent 402456 (1933) Improvements in gear driving train mountings particularly for electrically driven clocks

British Patent 419767 (1934) Improvements in or relating to starting devices for synchronous electric motors

British Patent 484222 (1938) Improvements relating to small synchronous electric motors

British Patent 502197 (1939) Improvements relating to electrically actuated alarm clocks

British Patent 521775 (1940) Improvements in the synchronisation of clocks

British Patent 5340439 (1940) Improvements relating to synchronous electric motors

British Patent 571849 (1945) Improvements in or relating to uni-directional drive mechanism

British Patent 744204 (1956) Improvements in or relating to the control of the rotational direction of synchronous electric motors

British Patent 806383 (1958) Improvements in methods of joining parts

Lines MA (2012) Ferranti synchronous electric clocks. Paperback edition with corrections. Zazzo Media, Milton Keynes

Miles RHA (2011) Synchronome. Masters of electrical time keeping. The Antiquarian Horological Society, Ticehurst

Pook LP (2014) Temco Art Deco domestic synchronous clocks. Watch Clock Bull 56/1(407): 47–58

Smith B (2008) Smiths domestic clocks, 2nd edn. Pierhead Publications Limited, Herne Bay

Chapter 9
Gallery of Synchronous Wall Clocks

Abstract A wall clock is a clock that is intended for display high up on a wall. In a domestic setting they are mostly used in kitchens. A wall clock has a dial that can be read from across the room. A time switch that controls central heating can be used as a wall clock which has a dial that can be read from a short distance. The main purpose of this gallery is to illustrate the range of British synchronous wall clocks that were available. There are back and front views of each clock, together with a brief description. Clocks are mostly illustrated as found, and some are in poor condition. A time switch is included. The sample of 16 synchronous wall clocks, and a time switch, illustrated is too small for general conclusions to be drawn. The large number of Smith clocks (41 %) is noticeable. A number of clocks (29 %) have Art Deco cases, but there are fewer clocks with other identifiable artistic styles. About half the clocks, and the time switch, (47 %) are self starting. None of them has an outage indicator. About half the clocks, and the time switch, (47 %) do not have any form of tell tale to make it easy for a user to determine whether or not a clock, or the time switch, is running. None of the clocks, or the time switch, show the date, or the day of the week.

9.1 Introduction

A wall clock is a clock that is intended for display high up on a wall. In a domestic setting they are mostly used in kitchens. A wall clock has a dial that can be read from across the room. A time switch that controls central heating can be used as a wall clock which has a dial that can be read from a short distance. The main purpose of this gallery is to illustrate the range of British synchronous wall clocks that were available. There are back and front views of each clock, together with a brief description. Clocks are mostly illustrated as found, and some are in poor condition. A time switch is included.

© Springer International Publishing Switzerland 2015   165
L.P. Pook, *British Domestic Synchronous Clocks 1930–1980*, History
of Mechanism and Machine Science 29, DOI 10.1007/978-3-319-14388-0_9

9.2   English Clock Systems Clock

English Clock Systems Synchronous Wall Clock   See Figs. 4.11 and 9.1. The clock is marked Made in Great Britain. It is for a 200–250 V supply. The style is Utility. The white metal case is 24.5 cm diameter. The unglazed dial is 21.5 cm diameter with black Arabic numerals and hands, and a red seconds hand. The brand on the front is English Clock Systems. The brand on the back is Smiths Industries Limited. The clock is fitted with a Smith self starting QGEM movement. The seconds hand acts as a tell tale.

9.3   Genalex Clock

Genalex Synchronous Wall Clock   See Fig. 9.2. This is a rebranded Smith *Durban*, date range 1937–1966 (Smith 2008). The clock is marked Made in England. It is for a 200–250 V supply. The style is Art Deco. The cream plastic case is 16.5 cm high × 16.5 cm wide. The dial is 11.5 cm diameter with a chrome bezel, cream Arabic numerals, black hour hand, and black and cream minute hand. The brand on the front and on the back is Genalex. The clock is fitted with a Smith Bijou movement. To start the clock press the starter knob and release. There is a tell tale in an aperture at 12 o'clock. British Patent information on the clock is '366710, 369438, 374713'.

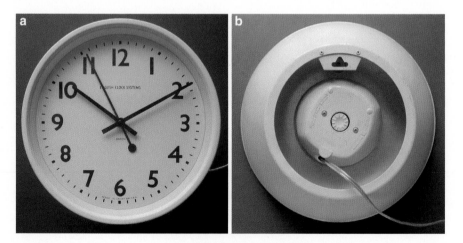

Fig. 9.1 English Clock Systems synchronous wall clock, (**a**) Front, (**b**) Back

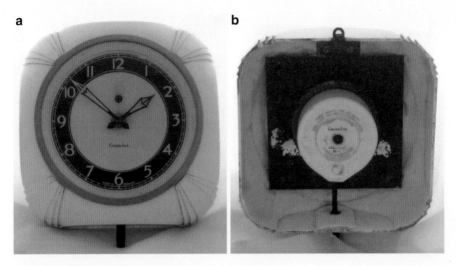

Fig. 9.2 Genalex synchronous wall clock, (**a**) Front, (**b**) Back

9.4 Horstmann Time Switch

Horstmann Type QMK3 Synchronous Time Switch See Fig. 9.3. The time switch is for a 200–250 V supply. The style is Utility. The grey and pale green plastic case is 18 cm high × 11 cm wide. The rotating dial is 8.5 cm diameter with black 24 h Arabic numerals, and a red line time indicator. The brand on the front is Potterton, a central heating boiler manufacturer. Brands on the front with the door open are Horstmann and Horstmann Gear Co Ltd. The time switch is fitted with a Horstman self starting movement. There is no tell tale. The movement plugs in for easy replacement. The central heating controls are simple and intuitive. British Patent information on the time switch is '824687'. This patent was issued in 1959.

9.5 Magneta Clock

Magneta Synchronous Wall Clock See Fig. 9.4. The clock is Made in England. It is for a 230 V supply. The chrome case is 22 cm diameter. The dial is 14 cm diameter with black Roman numerals and hands, and a red seconds hand. The brand on the front is Magneta. The brands on the back are Magneta and Magneta Time Co. Ltd. The clock is fitted with a Magneta Early movement. To start the clock pull the starter wire and release. The seconds hand acts as a tell tale.

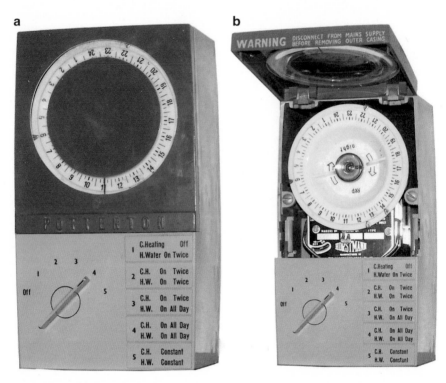

Fig. 9.3 Horstmann Type QMK3 Synchronous Time Switch (**a**) Front view, (**b**) Front view with door open

Fig. 9.4 Magneta synchronous wall clock, (**a**) Front, (**b**) Back

9.6   Metalair Clock

Metalair Synchronous Wall Clock See Fig. 9.5. The clock is marked Made in England. It is for a 200–250 V supply. The cream metal case is 18 cm diameter. The dial is 10 cm diameter with black Arabic numerals and hands. Brands on the front are Metalair and Metalair Ltd. The brand on the back is Metalair. A sticker on the back has the serial number 36287. The clock is fitted with a Metalair movement. To start the clock press the starter knob and release. The seconds hand acts as a tell tale.

9.7   Metamec Clock

Metamec Model 5635 Synchronous Wall Clock See Fig. 9.6. The date range is 1979–1983 (Bird 2003). The clock is marked Made in England. It is for a 220–240 V supply. The gilt metal case is 20.5 cm diameter. The dial is 19.5 cm diameter with gilt Arabic numerals, black hands, and a red seconds hand. The brand on the front and on the back is Metamec. The clock is fitted with a Metamec self starting SS 101 movement. The seconds hand acts as a tell tale.

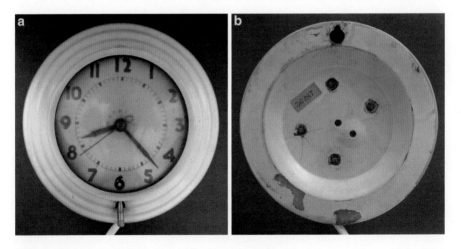

Fig. 9.5 Metalair synchronous wall clock, (**a**) Front, (**b**) Back

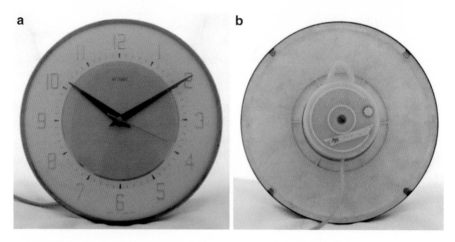

Fig. 9.6 Metamec Model 5635 synchronous wall clock, (**a**) Front, (**b**) Back

9.8   Smith Clock, Type 1 Movement

Smith *Derbyshire* Synchronous Wall Clock   See Fig. 9.7. The date is 1933 (Smith 2008). The clock is marked Made in England. It is for a 200–250 V supply. The style is Art Deco. The Bakelite case is 21.5 cm high × 21.5 cm wide. The dial is 15 cm diameter with black Arabic numerals and hands. There is a seconds indicator in an aperture at 12 o'clock. The brand on the front is Smith. Brands on the back are SEC, Smith's English Clocks Ltd and Synchronous Electric Clocks Ltd. The clock is fitted with a Smith Type 1 movement. To start the clock press the starter knob and release. The seconds indicator acts as a tell tale. British patent information on the clock is '366710, 369438, 374713, others pending'.

9.9   Smith Clock, BM7 Movement

Smith *Sussex* Synchronous Wall Clock   See Fig. 9.8. The date is 1963 (Smith 2008). The clock is marked Made in England. It is for a 200–250 V supply. The style is G-Plan. The wood veneer case is 31.5 cm diameter. The unglazed dial is 30.5 cm diameter with gilt Arabic numerals and hands. Brands on the back are SEC, Smith's English Clocks Ltd, and Synchronous Electric Clocks Ltd. The clock is fitted with a Smith BM7 movement. The movement cover projects 6.5 cm behind the keyhole hanging slot so a wall recess is required. To start the clock press the starter knob and release. There is no tell tale. British Patent information on the clock is '125953, 369438, 374713'.

Fig. 9.7   Smith *Derbyshire* synchronous wall clock, (**a**) Front, (**b**) Back

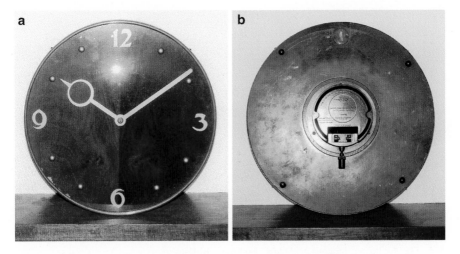

Fig. 9.8   Smith *Sussex* synchronous electric wall clock, (**a**) Front, (**b**) Back

9.10   Smith Clocks, BM39 Movement

Smith *Delhi de luxe* Synchronous Wall Clock   See Fig. 9.9. The date range is
1937–1939 (Smith 2008). The clock is marked Made in England. It is for a 200–
250 V supply. The black metal case is 40.5 cm diameter. The dial is 29 cm diameter
with black Arabic numerals and hands. The brand on the front is Smith. Brands on
the back are SEC, Smith's English Clocks Ltd, and Synchronous Electric Clocks

Fig. 9.9 Smith *Delhi de luxe* synchronous wall clock, (**a**) Front, (**b**) Back

Fig. 9.10 Smith synchronous wall clock, (**a**) Front, (**b**) Back

Ltd. The clock is fitted with a Smith BM39 movement. The front plate is marked 'BM39-36'. '36' is probably the year of manufacture, 1936. To start the clock press the starter knob and release. There is a tell tale in an aperture at 12 o'clock. British Patent information on the clock is '366710, 369438, 374713, others pending'.

Smith Synchronous Wall Clock See Fig. 9.10. The clock is marked Made in England. It is for a 200–250 V supply. The style is Art Deco. The pine and walnut case is 37 cm square. The unglazed dial is 32 cm square with white batons and hands. The clock is fitted with a Smith BM39 movement. Markings on the back plate include 'BM39' and '1 47'. The latter is probably the date of manufacture, January 1947. To start the clock press the starter knob and release. There is no tell tale. British Patent information on the clock is '366710, 369438, 374713, others pending'.

9.11 Smith Clock, Heavy Motion Work Movement

Smith *Oakley* Synchronous Wall Clock See Fig. 9.11. The date range is 1935–1937 (Smith 2008). The clock is marked Made in England. It is for a 200–250 V supply. The style is Art Deco. The wood veneer case is 23 cm high × 30.5 cm wide. The unglazed dial is 20.5 cm high × 28 cm wide with gilt Arabic numerals and hands. Brands on the back are SEC, Smiths English Clocks Ltd, and Synchronous Electric Clocks Ltd. The clock is fitted with a Smith self starting Heavy Motion Work movement, serial number 35858. There is no tell tale. British Patent information on the clock is '366710, 369438, 374713, others pending'.

9.12 Smith Clocks, Bijou Movement

Smith *Bounty* Synchronous Wall Clock See Fig. 9.12. The date range is 1954–1955 (Smith 2008). The clock is marked Made in Great Britain on the front and Made in England on the back. It is for a 200–250 V supply. The style is Pictorial. The painted metal case is 25.5 cm diameter. The unglazed dial is 16.5 cm diameter with black Arabic numerals and hands. Brands on the back are SEC and Smiths

Fig. 9.11 Smith *Oakley* synchronous wall clock, (**a**) Front, (**b**) Back

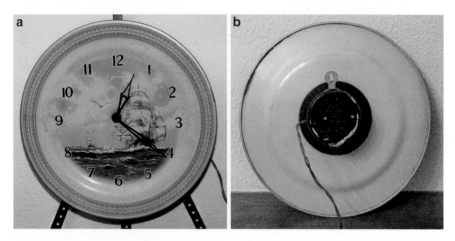

Fig. 9.12 Smith *Bounty* synchronous wall clock, (**a**) Front, (**b**) Back

Fig. 9.13 Smith *Durban* synchronous wall clock, (**a**) Front, (**b**) Back

English Clocks Ltd. The clock is fitted with a Smith self starting Bijou movement. There is no tell tale. British patent information on the clock is '366710, 369438, 374713, 384441, 484222'.

Smith *Durban* Synchronous Wall Clock See Fig. 9.13. The date range is 1937–1966 (Smith 2008). The clock is marked Made in Great Britain on the front and Made in England on the back. It is for a 200–250 V supply. The style is Art Deco. The blue plastic case is 16.5 cm high × 16.5 cm wide. The dial is 11.5 cm diameter with a chrome bezel, back Arabic numerals and hands, and a red seconds hand. Brands on the front are Smith and SECTRIC. Brands on the back are Smiths English Clocks, and SEC. The clock is fitted with a Smith Bijou movement. To start the clock press the starter knob and release. The seconds hand acts as a tell tale. British Patent information on the clock is '366710, 369438, 374713, 384441, 484222'. At the time of writing the clock had been in continuous use since 1990, with only routine resetting and restarting.

9.13 Synchronome Clocks

Synchronome Synchronous Wall Clock See Fig. 9.14. For more information see Miles (2011). A plaque on the back is marked 'E II R 1958' which suggests that the clock was made in 1958 for a UK government department. The clock is marked Made in England. It is for a 200–250 V supply. The style is Utility. The grey metal case is 28 cm diameter. The dial is 21 cm diameter with black Arabic numerals and hands. Brands on the back are Synchronome and Synchronome Co Ltd. The clock is fitted with a Smith Bijou movement branded Synchronome. To start the clock press the starter knob and release. There is no tell tale.

Synchronome Synchronous Wall Clock See Fig. 9.15. For more information see Miles (2011). The clock is marked Made in England on the front and Made in Great Britain on the back. It is for a 200–250 V supply. The style is Utility. The grey metal case is 28 cm diameter. The dial is 21.5 cm diameter with black Arabic numerals and hour hand, and a black and white minute hand. The brand on the

Fig. 9.14 Synchronome synchronous wall clock, (**a**) Front, (**b**) Back

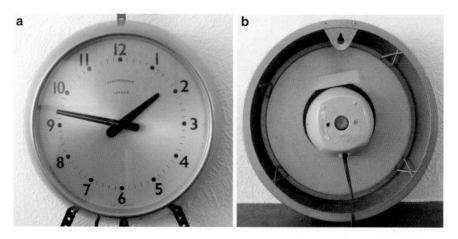

Fig. 9.15 Synchronome synchronous wall clock, (**a**) Front, (**b**) Back

front is Synchronome. Brands on the back are Synchronome and S Smith and Sons (England) Ltd. The clock is fitted with a Smith self starting QGEM movement. There is no tell tale. British Patent Information on the clock is '744204, 806383'.

9.14   Temco Clock, Mark V Movement

Temco Synchronous Wall Clock   See Fig. 9.16. The clock is marked Made in England. It is for a 200–250 V supply. The cream metal case is 27.5 cm diameter. The dial is 19 cm diameter with black Arabic numerals and hands. The brand on the front is Temco. Brands on the back are Temco and Telephone Mfg. Co. Ltd. The clock is fitted with a Temco bottom set Mark V movement. To start the clock press the starter knob and release. There is a tell tale in an aperture at 12 o'clock.

9.15   Westclox Clock

Westclox Synchronous Wall Clock   See Fig. 9.17. The clock is marked Made in Scotland. It is for a 200–250 V supply. The white and red plastic case is 21 cm high × 21 cm wide. The dial is 17 cm diameter with a chrome bezel, black Arabic numerals and hands, and a red seconds hand. The brand on the front is Westclox. The clock is fitted with a Westclox self starting BM25 movement. The seconds hand acts as a tell tale.

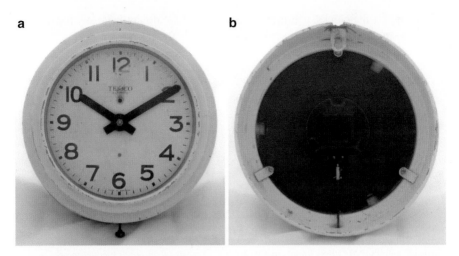

Fig. 9.16 Temco synchronous wall clock, (**a**) Front, (**b**) Back

Fig. 9.17 Westclox synchronous wall clock, (**a**) Front, (**b**) Back

9.16 Conclusions

1. The sample of synchronous wall clocks, and a time switch, illustrated is too small for general conclusions to be drawn.
2. The large number of Smith clocks (41 %) is noticeable.
3. A number of clocks (29 %) have Art Deco cases, but there are fewer clocks with other identifiable artistic styles.
4. About half the clocks, and the time switch, (47 %) are self starting. None of them has an outage indicator.

5. About half the clocks, and the time switch, (47 %) do not have any form of tell tale to make it easy for a user to determine whether or not a clock, or the time switch, is running.
6. None of the clocks, or the time switch, show the date, or the day of the week.

References

Bird C (2003) Metamec. The clockmaker. Dereham. Antiquarian Horological Society, Ticehurst
British Patent 125953 (1920) Improvements in and relating to electric time indicating apparatus
British Patent 366710 (1932) Improvements relating to electric motors
British Patent 369438 (1932) Improvements relating to electric clocks
British Patent 374713 (1932) Improvements relating to electric clocks
British Patent 384441 (1932) Improvements relating to electric clocks
British Patent 413119 (1934) Synchronous-motor clock
British Patent 484222 (1938) Improvements relating to small synchronous electric motors
British Patent 744204 (1956) Improvements in or relating to the control of the rotational direction of synchronous electric motors
British Patent 806383 (1958) Improvements in methods of joining parts
British Patent 824687 (1959) Improvements in electric time switches
Miles RHA (2011) Synchronome. Masters of electrical time keeping. The Antiquarian Horological Society, Ticehurst
Smith B (2008) Smiths domestic clocks, 2nd edn. Pierhead Publications Limited, Herne Bay

Chapter 10
Gallery of Synchronous Granddaughter Clocks

Abstract A granddaughter clock is a longcase clock that is less than 157 cm tall. A granddaughter clock has a dial that can be read from across the room. They became popular in the 1930s. In a domestic setting granddaughter clocks are mostly used in reception rooms and in entrance halls. This is the least common type of case used for domestic synchronous clocks. The main purpose of this gallery is to illustrate a few of the British synchronous granddaughter clocks that were available. There are back and front views of each clock, together with a brief description. Three of the four clocks are Smith clocks. Three of the four clocks are have Art Deco cases. None of the clocks are self starting. None of them has an outage indicator. Only one of the clocks has a tell tale to make it easy for a user to determine whether or not it is running. None of the clocks show the date, or the day of the week.

10.1 Introduction

A granddaughter clock is a longcase clock that is less than 157 cm tall. A granddaughter has a dial that can be read from across the room. They became popular in the 1930s. In a domestic setting, they are mostly used in reception rooms and in entrance halls. This is the least common type of case used for synchronous clocks. The main purpose of this gallery is to illustrate a few of the British synchronous granddaughter clocks that were available. There are back and front views of each clock, together with a brief description. The sample of granddaughter clocks illustrated is too small for conclusions to be drawn.

10.2 Ferranti Clock

Ferranti Model No. 151 Synchronous Granddaughter Clock See Figs. 4.1d and 10.1. The date range is 1936–1938 (Lines (2012). The clock is marked Made in England. It is for a 200–250 V supply. The style is Art Deco. The walnut veneer case is 142 cm high × 30.5 cm wide. The dial is 14.5 cm square with a chrome bezel, black Roman numerals, and chrome hands. The brand on the front and on the back, with the door open, is Ferranti. The back of the case is marked '36' probably

© Springer International Publishing Switzerland 2015
L.P. Pook, *British Domestic Synchronous Clocks 1930–1980*, History
of Mechanism and Machine Science 29, DOI 10.1007/978-3-319-14388-0_10

Fig. 10.1 Ferranti Model
No. 151 synchronous
granddaughter clock, (**a**)
Front, (**b**) Back

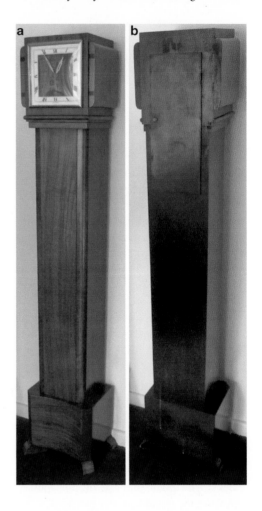

the year of manufacture, 1936. The clock is fitted with a Ferranti Early movement. To start the clock rotate the starter knob. The short seconds hand and rotating starter knob act as tell tales. British Patent information on the clock is '379383, 384681'.

10.3   Smith Clock, Type 1 Chiming Movement

Smith *Gibraltar* Synchronous Granddaughter Clock   See Figs. 1.4 and 10.2. The date range is 1935–1939 (Smith 2008). The clock is marked British Made. It is for a 200–250 V supply. The style is Art Deco. The walnut veneer case is 137 cm high × 25.5 cm wide. The original instructions are on the inside of the door. There is a cast iron weight in the base for stability. The dial is 16.5 cm square with a chrome bezel, black Roman numerals, and white hands. The brand on the front is Smith. The

Fig. 10.2 Smith *Gibraltar* synchronous granddaughter clock, (**a**) Front, (**b**) Back, door open

brand on the back, with the door open, is SEC. The clock is fitted with a Smith Type 1 chiming movement, serial number 3797. To start the clock push the starter lever and release. British Patent information on the clock is '366710, 369438, 374713'.

10.4   Smith Clock, Type 1 Movement

Smith Gibraltar Synchronous Granddaughter Clock   See Fig. 10.3. The date range is 1935–1939 (Smith 2008). The clock is marked British Made on the front, and Made in England on the back, with the door open. It is for a 200–250 V supply. The style is Art Deco. The oak veneer case is 137 cm high × 25.5 cm wide. The dial is 16.5 cm square with a chrome bezel, black Roman numerals, and white hands. The brand on the front is Smith. Brands on the back, with the door open, are SEC, Smith's English Clocks Ltd, and Synchronous Electric Clocks Ltd. The clock is fitted with a heavy motion work version of the Smith Type 1 movement. To start the clock press the starter knob and release. There is no tell tale. British Patent information on the clock is '366710, 369438, 374713, others pending'.

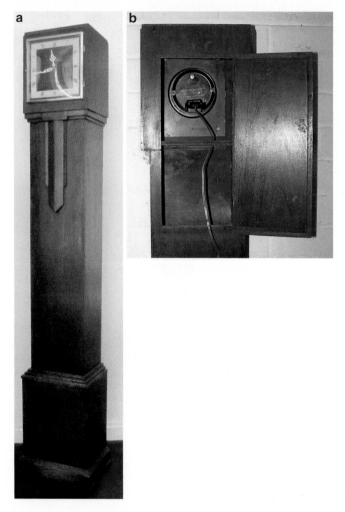

Fig. 10.3 Smith *Gibraltar* synchronous granddaughter clock, (**a**) Front, (**b**) Back, door open

10.5 Temco Clock, Mark IV Movement

Temco Synchronous Granddaughter Clock See Fig. 10.4. The clock is marked Made in England. It is for a 200–250 V supply. The walnut veneer case is 132 cm high × 30 cm wide. The dial is 14 cm diameter with a chrome bezel, and aluminium Arabic numerals and hands. Brands on the back, with the door open, are Temco, and Telephone Manufacturing Co. Ltd. The hand writing on the label above the door is illegible. The clock is fitted with a Temco Mark IV movement, serial number 84347. To start the clock rotate the starter knob. There is a tell tale in an aperture at 12 o'clock.

Fig. 10.4 Temco
synchronous granddaughter
clock, (**a**) Front, (**b**) Back

10.6 Conclusions

1. Three of the four clocks are Smith clocks.
2. Three of the four clocks are have Art Deco cases.
3. None of the clocks are self starting. None of them has an outage indicator.
4. Only one of the clocks has a tell tale to make it easy for a user to determine whether or not it is running.
5. None of the clocks show the date, or the day of the week.

References

British Patent 366710 (1932) Improvements relating to electric motors
British Patent 369438 (1932) Improvements relating to electric clocks
British Patent 374713 (1932) Improvements relating to electric clocks

British Patent 379383 (1932) Improvements in and relating to synchronous electric motors
British Patent 384681 (1932) Improvements in and relating to synchronous electric motors
Lines MA (2012) Ferranti synchronous electric clocks. Paperback edition with corrections. Zazzo
    Media, Milton Keynes
Smith B (2008) Smiths domestic clocks, 2nd edn. Pierhead Publications Limited, Herne Bay

Chapter 11
Gallery of Synchronous Movements

Abstract The purpose of this gallery is to illustrate synchronous movements, mostly from the British domestic synchronous clocks illustrated in the Chaps. 7, 8, 9, and 10, and to provide information on individual movements. All the movements were intended for use on the UK 50 Hz AC mains supply, and are mostly illustrated as found. Movement covers are usually moulded from Bakelite or plastic. The generic information in Chap. 3 on how synchronous movements work is suitable for most of the movements illustrated. The exception is the Marigold movement and explanations are included in its description. Most manufacturers made only a small number of different synchronous movements, with some variations to suit different clocks. The exception is Smith who made at least 13 different movements, together with a range of variations to suit different clocks. Both reluctance motors and magnetised motors can be satisfactory in service. The rotation speed of motors was a compromise. The most popular was 200 rpm, but this was not universal. Synchronous motors are not inherently self starting. Many clocks had manual start motors. Some had self starting motors, but reliability of self starting methods was sometimes a problem. Two types of reduction gear were used. In one type the reduction gear is similarly to the going train of a conventional mechanical clock. In the other type one, two or three worm gears were used. Both types can be satisfactory in service.

11.1 Introduction

The purpose of this gallery is to illustrate synchronous movements, mostly from the British domestic synchronous clocks illustrated in the Chaps. 7, 8, 9, and 10, and to provide information on individual movements. All the movements were intended for use on the UK 50 Hz AC mains supply, and are mostly illustrated as found. Movement covers are usually moulded from Bakelite or plastic. The generic information in Chap. 3 on how synchronous movements work is suitable for most of the movements illustrated. The exception is the Marigold movement and explanations are included in its description.

The usual method of supplying power to a movement is via terminals for a mains lead. Some synchronous movements are supplied via a 2 pin connector based on what used to be a British Standard 2 pin 2 A plug. The pin spacing is 12 mm.

© Springer International Publishing Switzerland 2015

L.P. Pook, *British Domestic Synchronous Clocks 1930–1980*, History of Mechanism and Machine Science 29, DOI 10.1007/978-3-319-14388-0_11

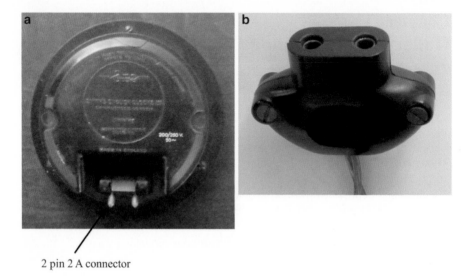

2 pin 2 A connector

Fig. 11.1 2 A 2 pin connector, (**a**) On a Smith Heavy Motion Work movement, (**b**) 2 pin 2 A female connector

The male connector on a Smith Heavy Motion Work movement (Sect. 11.13.4) is shown in Fig. 11.1a and a female connector in Fig. 11.1b. The female connector is sometimes missing from clocks and new replacements are not available. The best currently available alternatives are insulated 4 mm banana sockets, but these need to be used with care. A non standard 2 pin connector is used on Garrard movements (Sect. 11.7).

Most movement names are as used by manufacturers. When these are not available names are coined. The date of a movement is not necessarily a reliable guide to the date of the clock in which it is installed. Occasionally, clocks are marriages. When clocks were sent to a manufacturer for servicing the movement was sometimes replaced by a later movement when spares for the original movement were no longer available (Bird 2003).

Where available, British Patent numbers are included in descriptions. Various styles are used to indicate forthcoming British Patents. These are all cited as patent(s) pending.

11.2 BEM Movement

The BEM movement has a Sangamo magnetised motor. This has 30 poles on the stator so the motor rotates at 200 rpm. Adjacent poles on the stator have opposite polarity and are interleaved with air gaps between them. The magnetised rotor has two groups of three poles. One group is N and the other S. The movement has a double worm reduction gear. It is self starting with a hand set knob on the back. Power is supplied via a 2 pin 2 A connector. The movement cover is 3½ in. (88.9 mm) diameter × 2¼ in. (57.2 mm) deep with a flat at the bottom to clear the connector.

The Sangamo motor is also known as the Sangamo-Weston Type G motor, and is described by Robinson (1940a). British Patent information on Sangamo motors is: '377696, 384880'. An unused BEM movement is known. Its Sangamo motor has stickers showing the serial number '10894' and the date '21 FEB 1947'.

A rear view of the cover of a BEM movement is shown in Fig. 11.2a. Moulded on markings on the movement cover include 'BEM', 'BRITISH ELECTRIC METERS L^{TD}.' and 'LONDON AND BANGOR N. WALES'. Figure 11.2b shows a front view and Fig. 11.2c a rear view. Markings on the motor include 'SANGAMO'. A side view is shown in Fig. 11.2d and a top view in Fig. 11.2e. The markings which look like poles are on a rotating aluminium cover attached to the rotor. Figure 11.2f shows a front view of the motor and Fig. 11.2g a front view with the rotor and the rotating cover removed. The interleaved poles are visible.

11.3   Clyde Movement

The Clyde movement has a reluctance motor. There are 50 poles on the rotor so it rotates at 120 rpm. The reduction gear is similar to the going train of a conventional mechanical clock. On the back there is a push and release starter lever and a hand set knob. The movement cover is 3⅞ inches (98.4mm) diameter × 2¹⁄₁₆ inches (52.4mm) deep. Also see Sect. 3.2.4.2.

A rear view of the cover for a Clyde movement is shown in Fig. 11.3a. Moulded on markings include 'Clyde' and 'MADE IN SCOTLAND'. Figure 11.3b shows a rear view of the movement, Fig. 11.3c a top view and Fig. 11.3d a side view.

11.4   Elco Movement

The Elco movement has a reluctance motor. There are 30 poles on the rotor so it rotates at 200 rpm. The reduction gear is similar to the going train of a conventional mechanical clock. On the back there is a combined push and release starter and hand set knob. The movement cover is 2-7/8 in. (73.0 mm) diameter × 1-3/4 in. (44.5 mm) deep.

A version described by Robinson (1940a) has a push and release starting lever and separate hand set knob.

A cover for an Elco movement is shown in Fig. 11.4a. Markings include 'Elco CLOCKS & WATCHES L^{TD} LONDON' and 'MADE IN GREAT BRITAIN'. Figure 11.4b shows a clock case used as a cover. Markings include 'Elco CLOCKS & WATCHES L^{TD} LONDON' and 'MADE IN GREAT BRITAIN'. A front view of

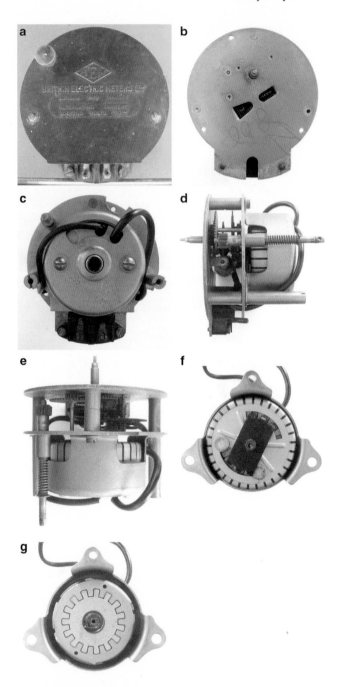

Fig. 11.2 BEM movement, (**a**) Rear view of cover, (**b**) Front view, (**c**) Rear view, (**d**) Side view, (**e**) Top view, (**f**) Front view of motor, (**g**) Front view of motor with rotor and rotating cover removed

Fig. 11.3 Clyde movement, (**a**) Cover, (**b**) Rear view, (**c**) Top view, (**d**) Side view

the movement is shown in Fig. 11.4c. The tell tale is near the top and the motion work is behind the front plate. Figure 11.4d shows a top view and Fig. 11.4e a rear view.

11.5   Empire Movements

The Empire movement has a reluctance motor. There are 30 poles on the rotor so it rotates at 200 rpm. There is a double worm reduction gear. The motor is unusual in that there are two coils on the stator. On the back there is a push and release starter lever and a hand set knob. The clock case is used as a movement cover. Power is supplied via a 2 pin 2 A connector.

Fig. 11.4 Elco movement, (**a**) Cover, (**b**) Clock case used as cover, (**c**) Front view, (**d**) Top view, (**e**) Rear view

A version of the movement used in Synchronomains clocks has a tell tale. The movement cover is 3-5/8 in. (92.1 mm) diameter × 1-3/8 in. (34.9 mm) deep.

The movement appears to have been introduced in 1932 (Miles 2011). British Patent information on Empire movements is: 'Patents pending' or 'Patent pending, 386756'. British Patent 386756 was issued in 1933, which tallies with a 1932 introduction date.

11.5.1 Empire Movement

A clock case used as a cover for an Empire movement is shown in Fig. 11.5a. Markings include 'EMPIRE' and 'MADE IN ENGLAND'. Figure 11.5b shows a rear view of the movement, and Fig. 11.5c a rear view with the back plate removed. A front view is shown in Fig. 11.5d. Markings on the front plate include 'EMPIRE', 'MADE IN ENGLAND' and the serial number 10120. There is an aperture for a tell tale, but no corresponding aperture in the clock dial. Figure 11.5e shows a front view of the motor removed from the movement. A view of the movement from underneath is shown in Fig. 11.5f.

11.5.2 Empire Movement for Synchronomains

A rear view of the cover of an Empire movement for Synchronomains is shown in Fig. 11.6a. Moulded on markings include 'SYNCHRONOMAINS', 'PATENT APPLIED FOR' and 'MADE IN ENGLAND'. Figure 11.6b shows a front view of the movement. Markings on the front plate include 'SYNCHRONOMAINS', 'MADE IN ENGLAND' and the serial number '12274'. A tell tale is visible through the aperture. A front view of the motor removed from the movement is shown in Fig. 11.6c. The red and white tell tale is attached to the motor shaft so rotates at 200 rpm and appears as a blur when the clock is running. Other views are as in Sect. 11.5.1.

11.6 Ferranti Movements

11.6.1 Ferranti Early Movement

The Ferranti Early movement has a reluctance motor. There are 36 poles on the rotor so it rotates at 166-2/3 rpm. The reduction gear is similar to the going train of a conventional mechanical clock. On the back there is a starter knob, which is rotated by hand, and a hand set knob. The movement cover is 3-5/8 in. (92.1 mm) diameter × 1-5/8 in. (41.3 mm) deep. The power consumption is slightly less than 1 W.

The Early movement was in use from 1931 to approximately 1945. It is described by Lines (2012), Philpott (1935), Robinson (1940a, 1942, 1946) and Wise (1951).

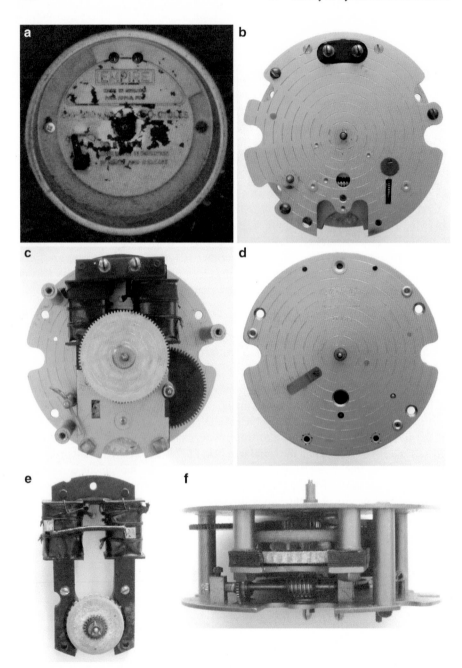

Fig. 11.5 Empire movement, (**a**) Clock case used as cover, (**b**) Rear view, (**c**) Rear view with back plate removed, (**d**) Front view, (**e**) Front view of motor, (**f**) View from underneath

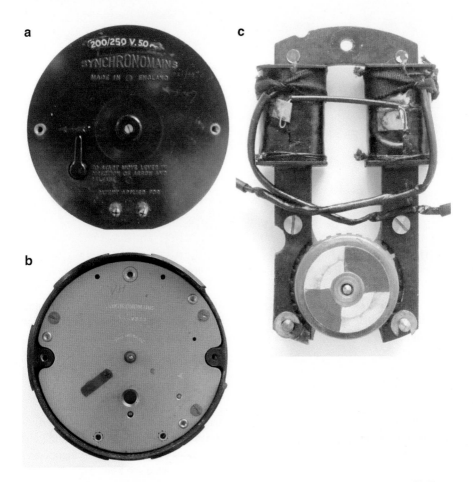

Fig. 11.6 Empire movement for Synchronomains, (**a**) Rear view of movement cover, (**b**) Front view of movement (**c**) Front view of motor

British Patent information on Early movements is: 'Patents pending' or '379383, 384681'. A prototype movement is reported as running at 142.8 rpm, which corresponds to 42 poles.

Several different cover styles were used for the Early movement (Lines 2012) and two of these are shown in Fig. 11.7. The cover shown in Fig. 11.7a is from a Ferranti Model No. 12 synchronous bedside clock (Fig. 8.5). Moulded on markings include 'MADE IN ENGLAND', 'FERRANTI' and patent application numbers '12942/31, 24461/31'. These can be interpreted as 'Patents pending'. The cover shown in Fig. 11.7b is from a Ferranti Model No. 132 synchronous bedside clock (Fig. 8.6). Moulded on markings include 'MADE IN ENGLAND' and 'FERRANTI'.

A front view of the movement is shown in Fig. 11.8a. The coil is visible below the wheels. The leads to the coil are permanently connected to the movement cover, so the coil has to be detached for removal of the rest of movement. Figure 11.8b

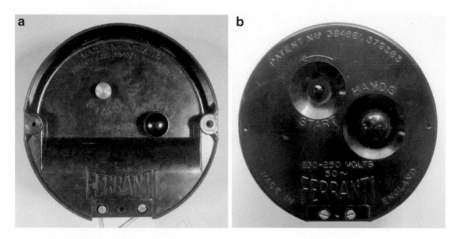

Fig. 11.7 Ferranti Early movement cover styles from, (**a**) Ferranti Model No. 12 synchronous bedside clock, (**b**) Ferranti Model No. 132 synchronous bedside clock

shows a front view of the movement without the coil, and Fig. 11.8c a front view with the front plate removed. Figure 11.8d shows a rear view of the movement with the coil removed, and Fig. 11.8e a view from underneath.

11.6.2 Ferranti Later Movement

The Ferranti Later movement has a reluctance motor. There are 30 poles on the rotor so it rotates at 200 rpm. There is a double worm reduction gear. On the back there is a starter knob, which is rotated by hand, and a hand set knob. The movement cover is 2½ in. (63.5 mm) diameter × 2 in. (50.8 mm) deep. Also see Sect. 3.2.4.1.

The Later movement was in use from approximately 1945 (Lines 2012). It was announced in 1947 and is described in detail by Robinson (1954). British Patent information is 'Patents pending'.

The cover used on Later movements is shown in Fig. 11.9a. Markings on the transfer include 'MADE IN ENGLAND' and 'FERRANTI'. Sometimes, the movement cover is unmarked and the transfer is on the back of the clock case. A transfer on the back of a clock case is shown in Fig. 11.9b. A rear view of the movement is shown in Fig. 11.10a, a side view in Fig. 11.10b and a top view in Fig. 11.10c.

11.7 Garrard Movements

Garrard movements have a reluctance motor. There are 30 poles on the rotor so it rotates at 200 rpm. Unusually, there is provision for oiling a rotor bearing. The coil can be removed by undoing two screws, hence a defective coil can easily be replaced. The reduction gear is similar to the going train of a conventional

Fig. 11.8 Ferranti Early movement, (**a**) Front view, (**b**) Front view, coil removed, (**c**) Front view, front plate removed, (**d**) Rear view, coil removed, (**e**) View from underneath

mechanical clock. There is a hand set knob on the back. Power is supplied via a non standard 2 pin connector. The plates are 3-7/8 in. (98.4 mm) diameter. They have a snailed and lacquered brass finish. The movement cover is 4 in. (101.6 mm) diameter × 1-3/4 in. (44.5 mm) deep. The coil has a separate cover. Also see Sect. 3.2.1.

British Patent information on Garrard movements is: 'Patent pending'.

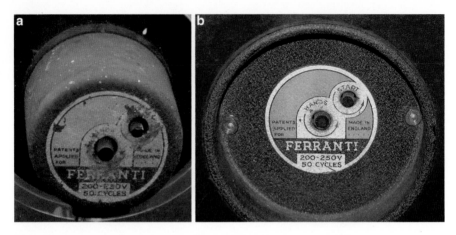

Fig. 11.9 Ferranti Later movement, (**a**) Cover, (**b**) Transfer on back of clock case

Fig. 11.10 Ferranti Later movement, (**a**) Rear view, (**b**) Side view, (**c**) Top view

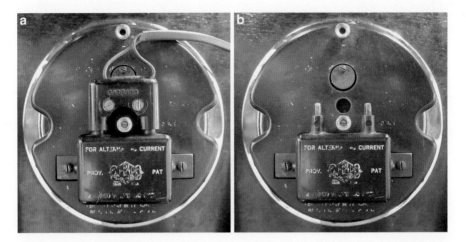

Fig. 11.11 Rear views of the cover of a Garrard Early movement, (**a**) Female connector in place, (**b**) Female connector removed

11.7.1 Garrard Early Movement

The Early movement has a twist and release starter knob on the back. It is described by Philpott (1935).

Rear views of the cover of an Early movement are shown in Fig. 11.11. Markings on the coil cover include 'MADE IN ENGLAND', 'PROV. PAT.' and 'GARRARD REGD <u>TRADE MARK</u>'. Figure 11.11a shows a rear view with the female connector in place. This is shaped to clear the hand set knob. A rear view with the female connector removed is shown in Fig. 11.11b. There is an aperture for a tell tale just above the hand set knob, and above this an access plate for oiling a rotor bearing.

A rear view of the Early movement with the coil removed is shown in Fig. 11.12a, and a rear view with a small plate removed in Fig. 11.12b. The contrasting paint on the rotor acts as a tell tale. It appears as a blur when the clock is running. A front view is shown in Fig. 11.12c. The front plate is marked 'GARRARD', 'MADE IN ENGLAND', and with the serial number '5162'. Figure 11.12d shows a side view and Fig. 11.12e a top view. The coil spring which operates the starter mechanism is visible.

11.7.2 Garrard Later Movement

The Garrard Later movement has a rotate to start knob on the back. It is described by Robinson (1940a).

A rear view of the cover a Later movement, with the coil in position is shown in Fig. 11.13a. Markings on the coil cover include 'MADE IN ENGLAND' and

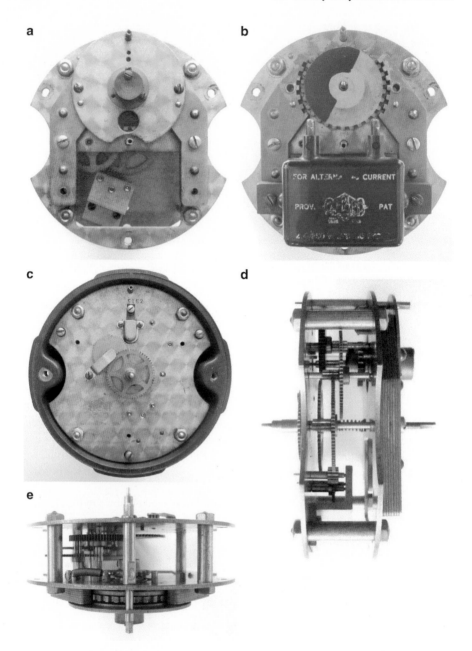

Fig. 11.12 Garrard Early movement, (**a**) Rear view, coil removed, (**b**) Rear view, small plate removed, (**c**) Front view, (**d**) Side view, (**e**) Top view

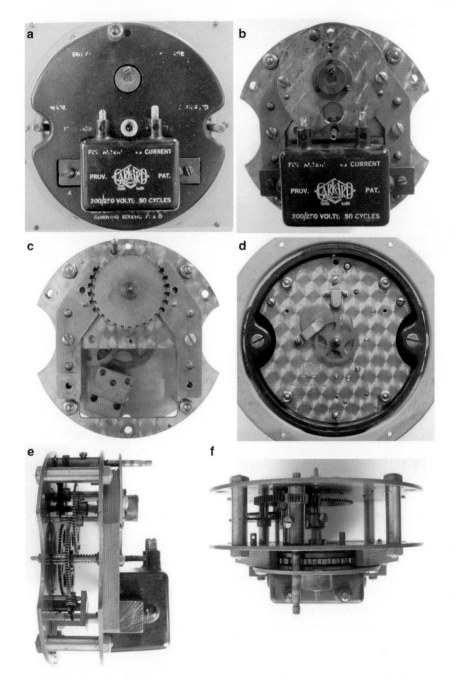

Fig. 11.13 Garrard Later movement, (**a**) Rear view of movement cover with coil in position, (**b**) Rear view, (**c**) Rear view with coil in position, (**d**) Front view, (**e**) Side view, (**f**) Top view

'GARRARD REGD <u>TRADE MARK</u>'. Figure 11.13b shows a rear view. There is an aperture for a tell tale just above the hand set knob, but there is no corresponding aperture in the movement cover. Above this an access plate for oiling a rotor bearing. Figure 11.13c shows a rear view with the coil and a small plate removed. A front view is shown in Fig. 11.13d. The front plate is marked 'GARRARD', 'MADE IN ENGLAND', and with the serial number '9403'. Figure 11.13e shows a side view and Fig. 11.13f a top view.

11.8 Goblin and Magneta Movements

Goblin and Magneta synchronous clocks were made in the same factory by the British Vacuum Cleaner & Engineering Co Ltd, Goblin Works, Leatherhead, Surrey (Anonymous 1939). The Goblin M.6. movement and Magneta Later movement are similar but with some detail differences.

11.8.1 Goblin M.6. Movement

The Goblin M.6. movement has a magnetised motor. There are 30 poles on the stator so the motor rotates at 200 rpm. Adjacent poles on the stator have opposite polarity and are interleaved with air gaps between them. The magnetised rotor has six pairs of poles. The pairs are alternately N and S. The reduction gear is similar to the going train of a conventional mechanical clock. The movement is self starting with a hand set knob at the back. The stator and rotor design is the same as that in the Smith Bijou movement and is used under license. Also see Sects. 3.2.2 and 3.2.4.3.

The M.6. movement is similar to the Magneta Later movement, but with some detail differences. The M.6. movement is described by Anonymous (1947). An earlier movement is described by Robinson (1940a).

A clock case used as the cover for an M.6. movement is shown in Fig. 11.14. Markings include 'GOBLIN', BRITISH MADE' 'PAT. No. 571849' and 'LICENSED UNDER LETTERS PAT. No. 366710'. This is a Smith patent. British Patent 571849, issued in 1945, is a British Vacuum Cleaner & Engineering Co Ltd patent. The movement could not have been made earlier than 1945.

A front view of the M.6. movement is shown in Fig. 11.15a. The motion work is behind the front plate. Figure 11.15b shows a rear view. A three-fourth rear view with the triangular plate removed is shown in Fig. 11.15c. The rotor, and the interleaved poles on the stator, are visible. Figure 11.15d shows a side view.

11.8.2 Magneta Early Movement

The Magneta Early movement has a reluctance motor. There are 30 poles on the rotor so it rotates at 300 rpm. The reduction gear is similar to the going train if a

Fig. 11.14 Goblin M.6.
movement, clock case used as
cover

conventional mechanical clock. There is a hand set knob on the back, and a pull and release starter wire underneath.

Figure 11.16a shows a rear view of an Early movement metal cover. Markings include 'MAGNETA TIME CO. LTD. LEATHERHEAD, SURREY.'. A rear view of the movement is shown in Fig. 11.16b, a top view in Fig. 11.16c, and a view from underneath in Fig. 11.16d.

11.8.3 Magneta Later Movement

The Later movement has a magnetised motor. There are 30 poles on the stator so the motor rotates at 200 rpm. Adjacent poles on the stator have opposite polarity, and are interleaved with air gaps between them. The magnetised rotor has six pairs of poles. The pairs are alternately N and S. The reduction gear is similar to the going train of a conventional mechanical clock. The movement is self starting with a hand set knob at the back. The rotor design is the same as that in the Smith Bijou movement, but the stator design is different. Also see Sect. 3.2.2.

The Later movement is similar to the Goblin M.6. movement, but with some detail differences.

British Patent information on the clock is '571849, other patents pending', British Patent 571849, issued in 1945, is a British Vacuum Cleaner & Engineering Co Ltd patent. The Later movement could not have been made earlier than 1945.

A clock case used as the cover for a Later movement is shown in Fig. 11.17a. Markings include 'MAGNETA TRADE MARK' and 'BRITISH MADE'. A rear view of the movement is shown in Fig. 11.17b and a rear view with one plate removed in Fig. 11.17c. The rotor and interleaved poles are visible. Unusually, a plate is held in place by wedges, two of which are visible near the top of the photograph. A rear view with two more plates removed is shown in Fig. 11.17d. A side view is shown in Fig. 11.17e and a front view in Fig. 11.17f.

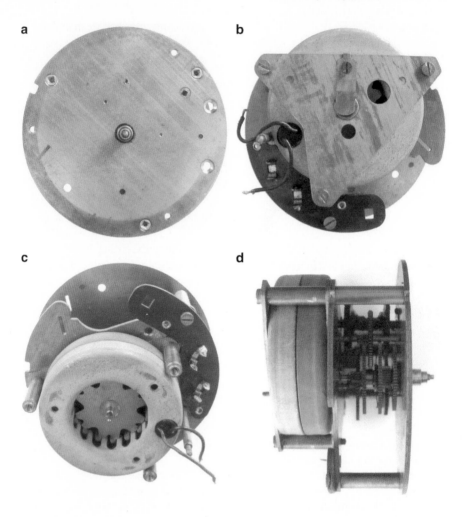

Fig. 11.15 Goblin M.6. movement, (**a**) Front view, (**b**) Rear view, (**c**) ¾ rear view, triangular plate removed, (**d**) Side view

The movement has an impulse self start. This is achieved by the uneven spacing of teeth on the rotor (Sect. 3.2.4.3) and a ratchet device to prevent rotation in the wrong direction. The three toothed ratchet wheel is visible in Fig. 11.17d. This engages with a pawl.

11.9   Kelnore Movement

The Kelnore movement has a reluctance motor. There are 30 poles on the rotor so it rotates at 200 rpm. The reduction gear is similar to the going train of a conventional mechanical clock. On the back there is a hand set knob and a rotate to start knob.

Fig. 11.16 Magneta Early movement, (**a**) Rear view of cover, (**b**) Rear view, (**c**) Top view, (**d**) View from underneath

A clock case used as a movement cover for a Kelnore movement is shown in Fig. 11.18a. Moulded on markings include 'J. H. JERRIM & C^O L^{TD} BIRMINGHAM'. Figure 11.18b shows a front view of the movement, Fig. 11.8c a rear view, and Fig. 11.18d a rear view with the back plate removed.

11.10 Marigold Movement

The Marigold movement has a reluctance motor. There are 50 poles on the rotor so it rotates at 120 rpm. The reluctance motor and reduction gear are unique designs that were not used by any other British synchronous clock manufacturers. The movement is not self starting. On the back there is a push and release starter lever and a hand set knob.

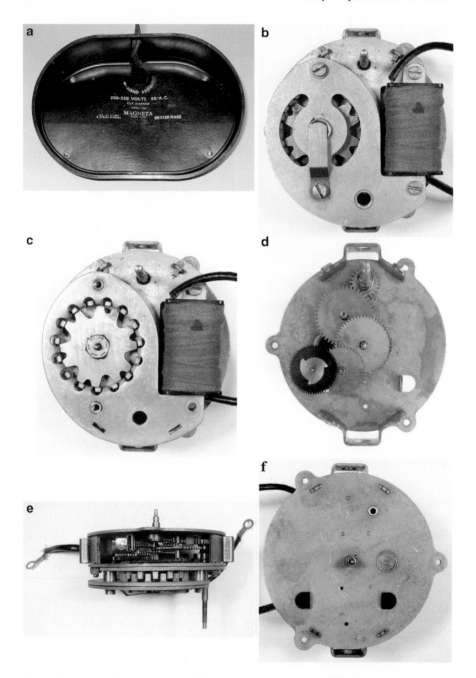

Fig. 11.17 Magneta Later movement, (**a**) Clock case used as cover, (**b**) Rear view, (**c**) Rear view, back plate removed, (**d**) Rear view, two more plates removed, (**e**) Side view, (**f**) Front view

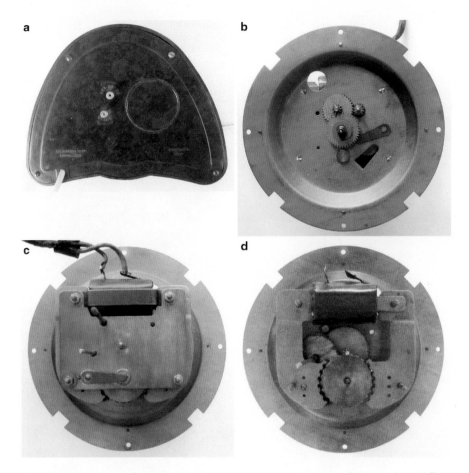

Fig. 11.18 Kelnore movement, (**a**) Clock case used as movement cover, (**b**) Front view, (**c**) Rear view, (**d**) Rear view with back plate removed

British Patent information is '444688'. The reduction gear used in the Marigold movement is described in Anonymous (1937a) and in British Patent 444688, which was issued in 1936. The movement illustrated in these references is different and has a magnetised motor. The patent date appears to date the movement to the late 1930s.

The main features of the Marigold movement are shown in Fig. 11.19. A front view is shown in Fig. 11.19a. The movement looks conventional with a starter lever, three wheels, a pinion, and a tell tale are visible. The tell tale rotates at 6 rpm. Figure 11.19b shows a three-fourth view of the movement. The stator is a hollow cylinder made from a magnetically soft material with the coil inside. One end of the

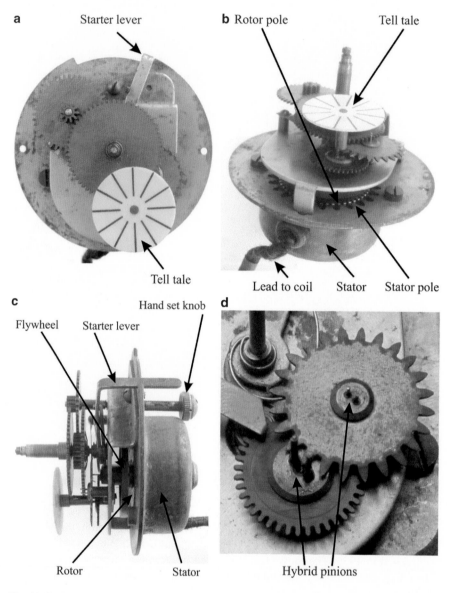

Fig. 11.19 Marigold movement, (**a**) Front view, (**b**) ¾ view, (**c**) Side view, (**d**) Detail view, partially dismantled

cylinder is extended so that the stator forms part of the clock case. There is a circular hole in this end with 25 poles pointing towards the centre line of the cylinder. Some of the stator poles and rotor poles are visible. A side view of the movement is shown in Fig. 11.19c. The flywheel smooths out the rotation of the rotor. The starter lever operates a conventional impulse starter.

The reduction gear is unconventional. Three of the wheels are driven by a hybrid pinion. A hybrid pinion (Fig. 11.19d) is a pair of rods mounted on a plate. Its form is that of a lantern pinion (Britten 1978), but with only two rods. Its action is like that of a Geneva stop mechanism, hence the title used by Anonymous (1937a). A Geneva stop mechanism has a single rod, and is sometimes used to prevent overwinding of a mainspring (Britten 1978). The wheel has teeth of the form used in lantern pinions, but cannot drive the hybrid pinion. The wheel is driven intermittently by the hybrid pinion. The use of hybrid pinions has the effect of making the lines on the tell tale jump across its aperture at intervals of slightly less than 1 s. The tell tale has 12 lines. If it had had 10 lines then it would have been a seconds indicator.

11.11 Metalair Movement

The Metalair movement has a reluctance motor. There are 30 poles on the rotor so it rotates at 200 rpm. The reduction gear is similar to the going train of a conventional mechanical clock. On the back there is a twist and release starter knob and a hand set knob. There was also a bottom set version with a combined starter and hand set knob.

Metalair started to manufacture synchronous clocks after the end of the Second World War in 1945. Clocks with guarantee cards dated April and May 1947 are known so the movement must have been in production at that time.

A rear view of a Metalair movement is shown in Fig. 11.20a, and rear view with the back plate and rotor removed in Fig. 11.20b. The movement has been smothered in what appears to be household oil in a misguided attempt to make it run. Figure 11.20c shows a top view.

11.12 Metamec Movements

There is relatively little information available on synchronous movements used in clocks made by Metamec (Miles 2003). Two of the synchronous movements made by Metamec are described. These are the Type 1 movement and the SS101 movement.

11.12.1 Metamec Type 1 Movement

The Type 1 movement has a reluctance motor. There are 30 poles on the rotor so it rotates at 200 rpm. The movement has a double worm reduction gear, and an optional artificial tick. On the back there is a turn past the click starter knob, a hand set knob, and a knob to turn the artificial tick on and off. The movement is built into the cover which is 3¼ in. (82.6 mm) diameter × 1½ in. (38.1) deep There is also

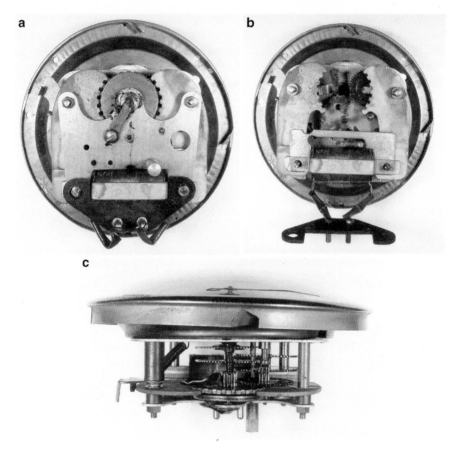

Fig. 11.20 Metalair movement, (**a**) Rear view, (**b**) Rear view, back plate and rotor removed, (**c**) Top view

a bottom set version with a combined hand set and push and release starter knob. There are self starting versions, an alarm version, and a version with a drive for an animation. Also see Sect. 3.2.1.

The Type 1 movement was first produced in late 1945. It is described in detail by Miles (2003) and Robinson (1956a). The artificial tick is sometimes regarded as a gimmick that was exclusive to Metamec, but is was popular with some people. An artificial tick was first used on a synchronous clock in 1930 in America, but was not popular Wemyss (1933).

Various cover styles were used for the Type 1 movement (Miles 2003) and two of these are shown in Fig. 11.21. The cover for a back set movement is shown in Fig. 11.21a. This has a flange for mounting the movement. The starter knob is on the left, the hand set knob in the centre, and the artificial tick control on the right. Moulded on markings on the cover include 'TURN START', 'METAMEC

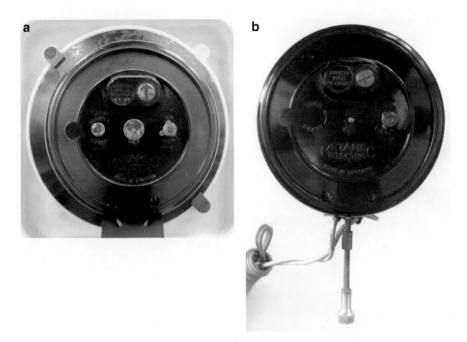

Fig. 11.21 Metamec Type 1 movement, **(a)** Cover for back set movement, **(b)** Cover for bottom set movement

ELECTRIC' and 'MADE IN ENGLAND'. The cover for a bottom set movement is shown in Fig. 11.21b. Moulded on markings are the same although 'TURN START' is redundant.

Figure 11.22a shows a front view of a back set movement, and Fig. 11.22b a front view with the front plate and rotor assembly removed. The rotor assembly is shown in Fig. 11.22c. The fibre wheel is an addition for an animation drive. A front view with the stator plate removed in Fig. 11.22d.

11.12.2 Metamec SS101 Movement

The SS101 movement has a magnetised motor. There are 30 poles on the rotor so it rotates at 200 rpm. The rotor consists of two discs, each with 15 poles bent over so that they are interleaved with air gaps between them and adjacent poles have opposite polarity. They are magnetised by a ferrite disc magnet between them. The movement has a double worm reduction gear, and extensive use is made of plastic parts. It is self starting with a hand set knob on the back. The movement cover is 3¹⁄₁₆ inches (77.8mm) diameter × 1³⁄₁₆ inches (30.2mm) deep.

Fig. 11.22 Metamec back set Type 1 movement, (**a**) Front view, (**b**) Front view with the front plate and rotor assembly removed, (**c**) Rotor assembly, (**d**) Front view with the stator plate removed

The SS101 movement is described by Miles (2003). He does not state when it was introduced. However, the SS101 movement is used in a clock with a date range of 1967–1993. This suggests a plausible introduction date of late 1960s.

The cover for an SS101 movement is shown in Fig. 11.23a. Moulded on markings include 'SS101', 'METAMEC' and 'MADE IN ENGLAND'. This example has a hanging loop for use in a wall clock, but this is not always provided. Figure 11.23b shows a front view and Fig. 11.23c a rear view. The interleaved poles on the rotor are visible. Figure 11.23d is a side view which shows the extensive use of plastic parts.

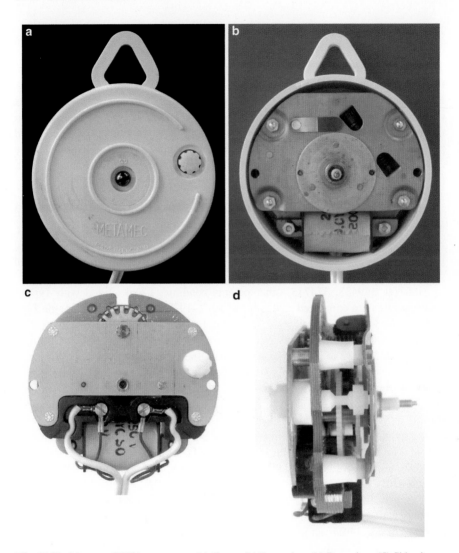

Fig. 11.23 Metamec SS101 movement, (**a**) Cover, (**b**) Front view, (**c**) Rear view, (**d**) Side view

11.13 Smith Movements

Smiths Group plc made a number of different types of synchronous movement. Some of these are described. Nomenclature is inconsistent and sometimes not available. Determining synchronous movement model names can be difficult. When model names are marked on movements these are used. Model names of unmarked movements can sometimes be determined by comparison with marked movements or from published sources. Otherwise, model names used are coined, sometimes by a previous author.

11.13.1 Smith Type 1 Movement

The Type 1 movement has a magnetised motor. There are 30 poles on the stator so the motor rotates at 200 rpm. Adjacent poles on the stator have opposite polarity and are interleaved with air gaps between them. The rotor is magnetised with six pairs of poles. The pairs are alternately N and S. The reduction gear is similar to the going train of a conventional mechanical clock. The movement has a combined press and release starter and hand set knob at the back. Power is supplied via a 2 pin 2 A connector. The movement cover is 3½ in. (88.9 mm) diameter × 2 in. (50.8 mm) deep. Bottom set and heavy motion work versions were also available. Also see Sect. 5.3.1

The Type I movement was introduced in 1931 (Smith 2008). The complicated magnetic design is described in detail by Ball (1932), Miles (2011) and Wise (1951). It was patented in 1932 (British Patent 366710). The same arrangement was used for most subsequent Smith synchronous movements. British Patent information given on Type 1 movements :is 'Patents pending', '366710, other patents pending' or '366710, 369438, 374713, other patents pending'. Clocks fitted with Type 1 movements have a combined date range of 1932–1933, which suggests that the Type I movement was in use until 1933. The patents were issued in 1932, which tallies with the clock dates. The heavy motion work version is used in a clock with date range of 1935–1939 so it was in use to at least 1935.

This type of motor is not self starting. It can be made self starting by the addition of an iron three legged spider (Miles 2011; Robinson 1942; Wise 1951). Starting in either direction is equally likely so a mechanical arrangement is needed to ensure starting in the correct direction.

A rear view of the movement cover of a Smith Type I movement is shown in Fig. 11.24a. Moulded on markings include 'SMITH SECTRIC', 'MADE IN ENGLAND', 'SMITH'S ENGLISH CLOCKS LTD' and 'Controlling SYNCHRONOUS ELECTRIC CLOCKS LTD CRICKLEWOOD, LONDON'. Figure 11.24b shows a front view of the movement, which is built into the movement cover. The seconds indicator is near the bottom. A front view with front plate and some wheels removed is shown in Fig. 11.24c and a front view with the back plate removed in Fig. 11.24d. The rotor and coil are visible. The interleaved poles, alternately N and S, are on the inside of the coil. The rear bearing of the rotor is moulded into the movement cover. The rotor assembly is shown in Fig. 11.24e. The rotor is sandwiched between two brass discs. Poles of the rotor are visible between the discs. The movement is not self starting. In the bottom set version the seconds indicator is at 12 o'clock.

A front view of the heavy motion work version is shown in Fig. 11.25. There is no seconds indicator.

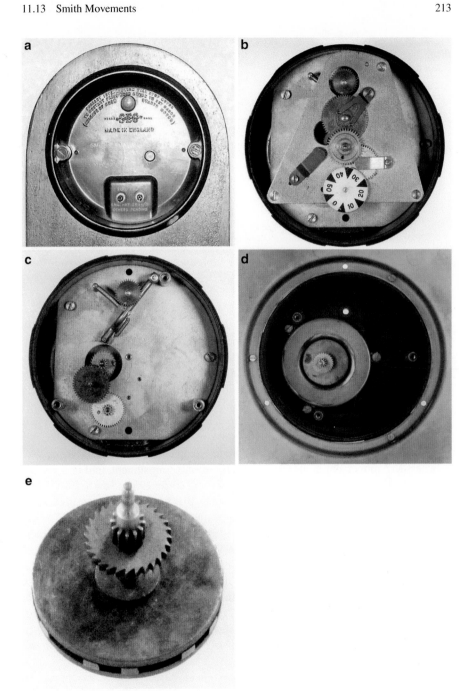

Fig. 11.24 Smith Type 1 movement, (**a**) Rear view of movement cover, (**b**) Front view,(**c**) Front view with front plate and some wheels removed, (**d**) Front view with the back plate removed, (**e**) Rotor assembly

Fig. 11.25 Front view of
Smith Type 1 movement,
heavy motion work version

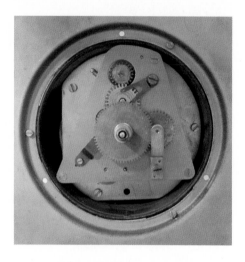

11.13.2 Smith BM7 Movement

The BM7 movement has a magnetised motor. There are 30 poles on the stator so
the motor rotates at 200 rpm. Adjacent poles on the stator have opposite polarity
and are interleaved with air gaps between them. The rotor is magnetised with six
pairs of poles. The pairs are alternately N and S. The movement has a double worm
reduction gear, and a combined press and release starter and hand set knob. Back set
and bottom set versions were available. Power is supplied via a swivelling 2 pin 2
A connector. The diameter of the plates is 3¼ in. (82.6 mm). The movement cover
is 3½ in. (88.9 mm) diameter × 2 in. (50.8 mm) deep.

The BM7 movement is also known as the de luxe movement, and is a revised
version of the Type 1 movement. It was introduced in 1934 (Smith 2008). It is
described by Robinson (1940a, 1942) who states that it was in production in 1939.
British Patent information on one BM7 movement is: '366710, 369438, 374713',
the date of the clock in which it is installed is unknown. On another BM7 movement
British Patent information is '125953, 369438, 374713', the date of the clock on
which it is installed is 1963. It is not clear why '125953' is on the movement
because the British Patent is not relevant to synchronous movements. The British
Patent information on several BM7 movements is '366710, 369438, 374713, other
patents pending'. The combined date range of the clocks in which these movements
are installed is 1932–1939. The clock dates do not quite tally with the date of
introduction of the BM7 movement, but it is clear that the movement was in use
until around 1963.

Figure 11.26 shows a back set BM7 movement. A rear view of the movement
cover is shown in Fig. 11.26a. Moulded on markings include 'TRADE SEC
MARK', 'SMITH'S ENGLISH CLOCKS L^TD CRICKLEWOOD LONDON

Fig. 11.26 Smith back set BM7 movement, (**a**) Rear view of movement cover, (**b**) Rear view of movement, (**c**) Top view of movement, (**d**) View from underneath, (**e**) Front view of movement

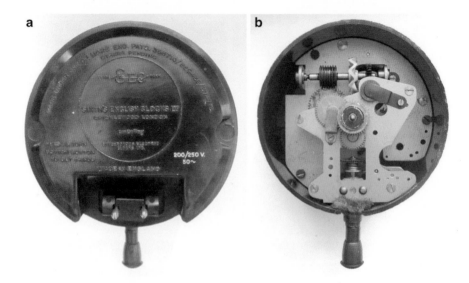

Fig. 11.27 Smith bottom set BM7 movement, (**a**) Rear view of movement cover, (**b**) Front view of movement

Controlling SYNCHRONOUS ELECTRIC CLOCKS LTD' and 'MADE IN ENGLAND'. Figure 11.26b shows a rear view of the movement and Fig. 11.26c a top view. The red wiggly line on the tell tale is hand painted over the original black line. A view from underneath is shown in Fig. 11.26d, and a front view of the movement in Fig. 11.26e. 'BM7-36' is stamped on the front plate.

Figure 11.27 shows a bottom set BM7 movement. A rear view of the movement cover is shown in Fig. 11.27a. Moulded on markings are the same as those on the back set version. Figure 11.27b shows a front view of the movement. The tell tale differs from that shown in Fig. 11.26c, and is as illustrated by Robinson (1942).

11.13.3 Smith BM39 Movement

The BM39 movement has a magnetised motor. There are 30 poles on the stator so the motor rotates at 200 rpm. Adjacent poles on the stator have opposite polarity and are interleaved with air gaps between them. The rotor is magnetised with six pairs of poles. The pairs are alternately N and S. The movement has a double worm reduction gear, and a combined press and release starter and hand set knob at the bottom. Power is supplied via a swivelling 2 pin 2 A connector. The diameter of the plates is 3¼ in. (82.6 mm). The movement cover is 3½ in. (88.9 mm) diameter × 2 in. (50.8 mm) deep. It does not appear to have been made in a back set version.

Fig. 11.28   Smith BM39 movement covers, (**a**) Moulded on markings, (**b**) Moulded on and printed markings

The BM39 movement is similar to the BM7 movement and it is not clear why it was given a separate designation. A BM39 movement with a serial number is known. Available dates suggest that the date range is at least 1936–1947. British patent information on BM39 movements is: '366710, 369438, 374713, other patents pending'.

Two types of movement cover are known. The moulded on markings on the cover shown in Fig. 11.28a are the same as those on a BM7 movement cover. The printed on markings on the cover shown in Fig. 11.28b include 'TRADE SEC MARK', 'SMITH'S ENGLISH CLOCKS L^TD', 'CRICKLEWOOD LONDON' and 'MADE IN ENGLAND'

Figure 11.29a shows a front view of the BM39 movement from the cover shown in Fig. 11.28a. 'BM39-36' is stamped on the front plate. '36' appears to be the date of manufacture, 1936. A front view of the movement from the cover shown in Fig. 11.28b, with the front plate removed, is shown in Fig. 11.29b. Markings on the front of the back plate include 'BM39' and '1 47' The latter appears to be the date of manufacture, January 1947.

11.13.4   Smith Heavy Motion Work Movement

The Heavy Motion Work movement has a magnetised motor. There are 30 poles on the stator so the motor rotates at 200 rpm. Adjacent poles on the stator have opposite

Fig. 11.29 Smith BM39 movement, (**a**) Front view, (**b**) Front view with front plate removed

polarity and are interleaved with air gaps between them. The rotor is magnetised with six pairs of poles. The pairs are alternately N and S. The movement has a double worm reduction gear. The movement is self starting, and there is no hand setting. A bottom set version was available. Power is supplied via a swivelling 2 pin 2 A connector. The diameter of the plates is 3¼ in. (82.6 mm). The movement cover is 3½ in. (88.9 mm) diameter × 2 in. (50.8 mm) deep.

This is a self starting, heavy motion work version of the BM7 movement. It is described by Robinson (1940a, 1942). It was in production in 1939. The bottom set version was in production in 1949. It was in use until around 1955 (Smith 2008) British Patent information on Heavy Motion Work movements is: 366710, 369438, 374713, other patents pending'.

Figure 11.30 shows a Heavy Motion Work movement. A rear view of the movement cover is shown in Fig. 11.30a. Moulded on markings are the same as those on the BM7 movement cover (Fig. 11.26a). A front view of the movement is shown in Fig. 11.30b. The large wheels in the motion work mean that there is no provision for a tell tale.

Figure 11.31 shows a bottom set Heavy Motion Work movement. A rear view of the movement cover is shown in Fig. 11.31a. Moulded on markings are the same as those on the BM7 movement (Fig. 11.26a). A front view of the movement is shown in Fig. 11.31b. The large wheels in the motion work mean that there is no provision for a tell tale. Figure 11.31c shows a front view with the front plate removed. The faint markings on the front of the back plate include the serial number '35858'. A rear view of the movement is shown in Fig. 11.31d, a top view in Fig. 11.31e, and a side view in Fig. 11.31f.

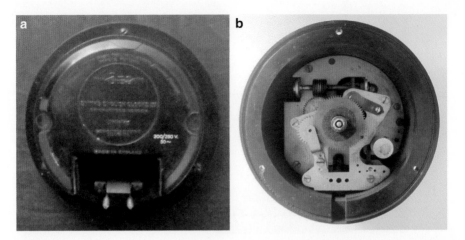

Fig. 11.30 Smith Heavy Motion Work movement, (**a**) Rear view of movement cover, (**b**) Front view of movement

11.13.5 Smith Bijou Movement

The Bijou movement has a magnetised motor. There are 30 poles on the stator so the motor rotates at 200 rpm. Adjacent poles on the stator have opposite polarity and are interleaved with air gaps between them. The rotor is magnetised with six pairs of poles. The pairs are alternately N and S. The reduction gear is similar to the going train of a conventional mechanical clock, with all wheels and pinions on stud mountings. The movement has a combined press and release starter and hand set knob either at the back or at the bottom. Another version has a twist and release starter knob, and a hand set knob, at the back, and there is also a self starting version. The diameter of the plates is 2-1/4 in. (57.2 mm). The movement cover is 2-5/8 in. (66.7 mm) across × 1-5/8 in. (41.3 mm) deep. The movement cover for the twist and release starter knob version is 1-7/8 in. (47.6 mm) deep.

As indicated by its name, the Bijou movement is smaller than the Type 1 and BM7 movements. It was introduced in 1937 (Anonymous 1937c). It came into general use in 1938, and was in use until the late 1950s (Smith 2008). It is described by Robinson (1940a, 1942) and Wise (1951) and in detail by Miles (2011). Detail changes made since its introduction are described by Robinson (1956b). It is usually very reliable. A clock with a Bijou movement is known to have been in continuous use for over 20 years without servicing. British Patent information on various Bijou movements is. (1) '366710, 369438, 374713' and the clock date range is 1937–1966'. (2) '366710, 369438, 374713, other patents pending', the clock date is unknown. (3) '366710, 374713, 384441, 484222', the clock date is unknown. (4) '366710, 369438, 374713, 384441, 484222', the combined clock date range is 1937–1955. (5) '366710, 369438, 374713, 384441, 484222, other patents pending',

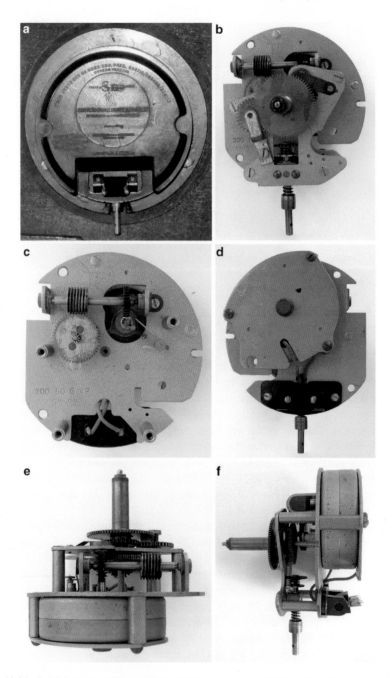

Fig. 11.31 Smith bottom set Heavy Motion Work movement, (**a**) Rear view of movement cover, (**b**) Front view of movement, (**c**) Front view with the front plate removed, (**d**) Rear view of movement, (**e**) Top view of movement, (**f**) Side view of movement

the clock date is 1950. Unbranded Bijou movements, and Bijou movements branded on the cover Ingersoll and Synchronome, are known.

Several different cover styles were used for the Bijou movement (Smith 2008) and some of these are shown in Fig. 11.32. Movement covers are usually, but not always, black. A movement cover for a bottom set movement is shown in Fig. 11.32a. Moulded on markings include 'MADE IN ENGLAND BY SMITHS ENGLISH CLOCKS LTD. LONDON' and 'TRADE SEC MARK'. Figure 11.32b shows a movement cover for a back set movement with the knob at bottom centre. Moulded on markings include 'MADE IN ENGLAND', 'SMITHS ENGLISH CLOCKS' and 'SEC REG^D TRADE MARK'. A movement cover for a back set movement with the knob at bottom right is shown in Fig. 11.32c. Moulded on markings include 'SEC REG^D TRADE MARK' and 'MADE IN ENGLAND'. Figure 11.32d shows a transparent movement cover for a self starting movement with no setting arrangement. The back of the movement is visible through the open clock door. The dial is unglazed so hand setting is by turning the hands. A movement cover for a back set movement with a twist and release starter knob is shown in Fig. 11.32e. Moulded on markings include 'MADE IN ENGLAND' but there is no maker's name. Sometimes the clock case is used as a movement cover. An example is shown in Fig. 11.32f. Markings on the back of the clock include 'SMITH SECTRIC' and 'MADE IN ENGLAND'.

A rear view of a Bijou movement is shown in Fig. 11.33a and a front view in Fig. 11.33b. A front view with the front plate removed is shown in Fig. 11.33c and a front view with some wheels removed in Fig. 11.33d. A top view is shown in Fig. 11.33e and a view from underneath in Fig. 11.33f.

A motor from a Smith Bijou movement is shown in Fig. 11.34a, and with the rotor removed in Fig. 11.34b. Adjacent poles on the stator have opposite polarity and are interleaved with air gaps between them. Figure 11.34c shows the rotor. This is magnetised with six pairs of poles. The pairs are alternately N and S.

11.13.6 Smith QAT Movement

The QAT movement has a reluctance motor. There are 30 poles on the rotor so it rotates at 200 rpm. The movement has a double worm reduction gear. There is a push and release starter lever, and a push in hand set knob. Movement covers are 3½ in. (88.9 mm) diameter.

The QAT movement was developed in 1953 (Smith 2008). It is unusual for a Smith movement in that it has a reluctance motor.

Figure 11.35 shows two types of cover used for QAT movements. That shown in Fig. 11.35a is used on a mantel clock and that in Fig. 11.35b on a wall clock. Moulded on markings include 'SMITHS ENGLISH CLOCKS LTD' and 'MADE IN ENGLAND'. The cover in Fig. 11.35b is marked 'B.S. 472'. This is the British Standard for synchronous clocks. It was first issued in 1932 and last revised in 1962 (Anonymous 1932, 1962). Figure 11.35c shows a clock case used as a movement

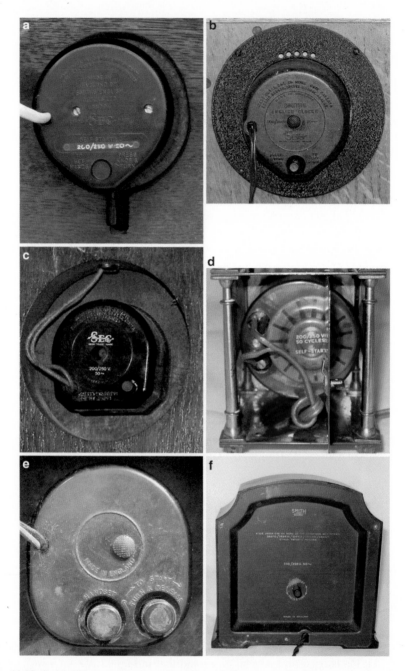

Fig. 11.32 Smith Bijou movement cover styles, (**a**) Bottom set (**b**) Back set, knob at bottom centre, (**c**) Back set, knob at bottom right, (**d**) No setting arrangement, (**e**) Back set, twist knob and release, (**f**) Clock case used as cover

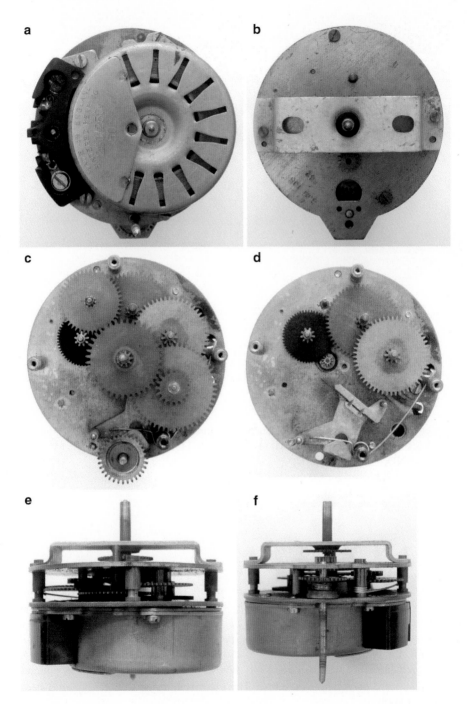

Fig. 11.33 Smith Bijou movement, (**a**) Rear view, (**b**) Front view, (**c**) Front view with front plate removed, (**d**) Front view with some wheels removed, (**e**) Top view, (**f**) View from underneath

Fig. 11.34 Motor from Smith Bijou movement, (**a**) Motor, (**b**) Motor with rotor removed, (**c**) Rotor

cover on a Smith *Huntingdon* Synchronous Mantel Clock (Fig. 7.44). Markings on the back of the clock include 'SMITHS ENGLISH CLOCKS LTD' and 'MADE IN ENGLAND'.

A rear view of the QAT movement is shown in Fig. 11.36a and a three-fourth side view in Fig. 11.36b.

11.13.7   Smith QEMG Movement

The QEMG movement has a magnetised motor. There are 30 poles on the stator so the rotor rotates at 200 rpm. Adjacent poles on the stator have opposite polarity and are interleaved with air gaps between them. The rotor is a plain ferrite disc magnetised with six pairs of poles. The pairs are alternately N and S. The motor is vertical with the rotor at the top. The movement is self starting, and has a single worm reduction gear. In the rest of the train steel wheels mesh with brass pinions. The plates are plastic bonded fabric. There are versions with and without a back hand set knob.

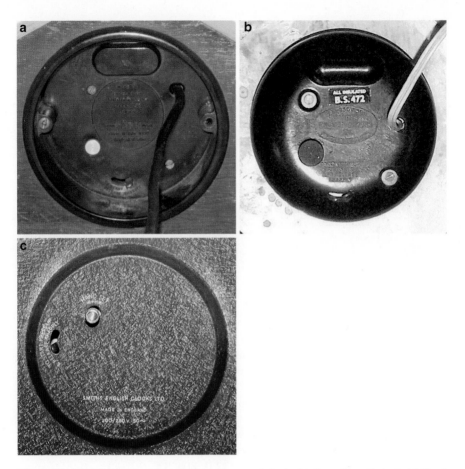

Fig. 11.35 Smith QAT movement, (**a**) Cover for mantel clock, (**b**) Cover for wall clock, (**c**) Clock case used as cover on a Smith *Huntingdon* Synchronous Mantel Clock

The QEMG movement was introduced in 1957, and was intended as a replacement for the Bijou movement (Miles 2011; Smith 2008). It is described in detail by Miles, who calls it the EMG movement. British Patent information on QEMG movements is: '744204, other patents pending'.

A cover for a QEMG movement is shown in Fig. 11.37a. Moulded on markings include 'SMITHS CLOCKS AND WATCHES LTD. MADE IN ENGLAND'. Although there is a hole for one there is no hand set knob. It is not needed because the dial is unglazed. Figure 11.37b shows a clock case used as a cover. Markings on the back of the clock include 'SMITHS CLOCKS & WATCHES LTD.' and 'MADE IN ENGLAND'. A rear view of the movement is shown in Fig. 11.37c and a side view in Fig. 11.37d.

Fig. 11.36 Smith QAT movement, (**a**) Rear view, (**b**) ¾ side view

11.13.8   Smith QGEM Movement

The QGEM movement has a magnetised rotor. There are 30 poles on the stator so the motor rotates at 200 rpm. Adjacent poles on the stator have opposite polarity and are interleaved with air gaps between them. The rotor is a plain ferrite disc magnetised with six pairs of poles. The pairs are alternately N and S. The motor is vertical with the rotor at the top. The movement is self starting, and has a double worm reduction gear. The plates and most of the wheels are plastic. Also see Sects. 3.2.2 and 5.3.1.

The QGEM movement was introduced in 1963, and was another attempt to produce cheaper movements by the use of precision plastic mouldings (Smith 2008). It is described in detail by Miles (2011) who calls it the GEM movement. Apart from slightly different poles the motor is the same as the QEMG movement. British Patent information on QGEM movements is: '744204, 806383'.

Two different cover styles were used for the QGEM movement. A round movement cover is shown in Fig. 11.38a. Moulded on markings include 'SMITHS CLOCKS AND WATCHES LTD.' and 'MADE IN GREAT BRITAIN'. The cover includes a hanger for a wall clock. In this particular clock the movement was mounted upside down. Figure 11.38b shows a square movement cover used where space was restricted. Moulded on markings include 'SMITHS INDUSTRIES LIMITED' and 'MADE IN GREAT BRITAIN'. Some QGEM covers are marked 'S. SMITH & SONS (ENGLAND) LTD.'

A rear view of the QGEM movement is shown in Fig. 11.39a. Markings on the back plate include 'S SMITH & SONS (ENGLAND) LTD' GREAT BRITAIN' and 'QGEM'. Figure 11.39b shows a three-fourth top view and Fig. 11.39c a view from underneath. In the latter figure the movement is in poor condition, with a damaged back plate and badly rusted mounting plate.

Fig. 11.37 Smith QEMG movement, (**a**) Cover, (**b**) Clock case used as cover, (**c**) Rear view, (**d**) Side view

11.13.9   Smith Type 1 Striking Movement

The Type 1 Striking movement has a magnetised motor. There are 30 poles on the stator so the motor rotates at 200 rpm. Adjacent poles on the stator have opposite polarity and are interleaved with air gaps between them. The rotor is magnetised with six pairs of poles. The pairs are alternately N and S. There is a push and release starter lever and a hand set knob. The train splits with a double worm reduction

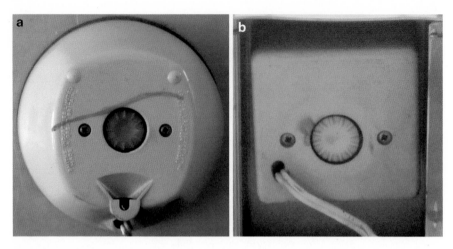

Fig. 11.38 Smith QGEM movement cover styles, (**a**) Round cover, (**b**) Square cover

gear to the motion work for the hands, and conventional wheels for the rack striking work. Striking is on a gong. The rack striking work is based on that used in Smith mechanical clocks (Smith 2008), except that it is driven via a jockey wheel which engages when striking takes place, rather than by a mainspring.

A sketch of the Type 1 Striking movement is reproduced by Smith (2008) who states that it is included in the 1933 catalogue. It is used in clocks with dates of 1933 and 1934. British Patent information on Type 1 Striking movements is: '366710, other patents pending'. Serial numbers 2626 and 6153 are known.

A rear view of the movement is shown in Fig. 11.40a. The terminal cover is missing. Striking is on a gong. The motor is completely enclosed. Markings on the back plate are 'SMITHS ENGLISH CLOCKS L^{TD}', CRICKLEWOOD LONDON', 'TRADE SEC MARK and the serial number '6153'. Figure 11.40b shows a front view of the movement. The rack striking work is visible. A front view of the movement with the front plate removed is shown in Fig. 11.40c. A top view is shown in Fig. 11.40d. The drive to the striking work is visible.

11.13.10   Smith Narrow Striking Movement

The Narrow Striking movement has a magnetised motor. There are 30 poles on the stator so the motor rotates at 200 rpm. Adjacent poles on the stator have opposite polarity and are interleaved with air gaps between them. The rotor is magnetised with six pairs of poles. The pairs are alternately N and S. There is a push and release starter lever and a hand set knob. The train splits with a double worm reduction gear to the motion work for the hands, and conventional wheels for the rack striking work. Striking is on a gong. A jockey wheel engages when striking takes place.

Fig. 11.39 Smith QGEM movement, (**a**) Rear view, (**b**) ¾ top view (**c**) View from underneath

Power is supplied via a 2 pin 2 A connector. The size over plates is 3½ inches (88.9 mm) high × 5¾ in. (146.1 mm) wide. The width between the plates is 14.5 mm and the overall depth, including the hammer and gong, is 70 mm. Also see Sects. 3. 4 and 5.3.1.

The Narrow Striking movement is described by Robinson (1940a) and Miles (2011). It was produced in 1935 for use in shallow case clocks, which were becoming fashionable (Anonymous, 1935, 1936). British Patent information on Narrow Striking movements is '387108'.

A rear view of the movement is shown in Fig. 11.41a. Striking is on a gong. The motor is completely enclosed. Markings on the back plate include 'SMITHS ENGLISH CLOCKS L^{TD.} CRICKLEWOOD LONDON'. This marking has faded and is barely visible. The movement has an easily removable spring steel cover. A top view with the cover in place is shown in Fig. 11.41b and with the cover removed

Fig. 11.40 Smith Type 1 Striking movement, (**a**) Rear view, (**b**) Front view, (**c**) Front view with front plate removed, (**d**) Top view

in Fig. 11.41c. A front view is shown in Fig. 11.41d. The rack striking work is visible. A rear view of the front plate with some wheels in position is shown in Fig. 11.41e. The red painted wheel acts as a tell tale. The wheel above it is the jockey wheel which engages when striking takes place. A rear view of the motor, removed from the movement is shown in Fig. 11.41f.

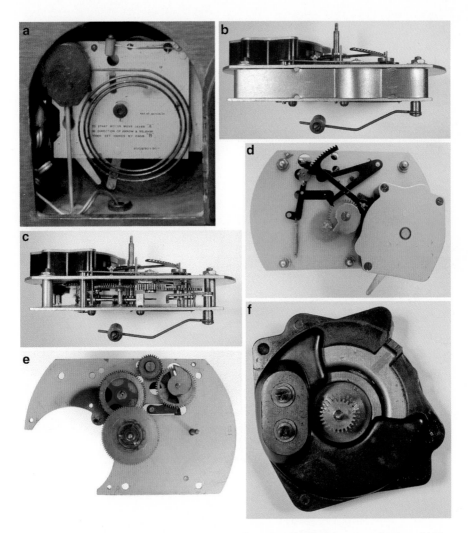

Fig. 11.41 Smith Narrow striking movement, (**a**) Rear view (**b**) Top view, cover in place, (**c**) Top view, cover removed, (**d**) Front view, (**e**) Rear view of front plate, (**f**) Rear view of motor

11.13.11 Smith Type 1 Chiming Movement

The Type 1 Chiming movement has a magnetised motor. There are 30 poles on the stator so the motor rotates at 200 rpm. Adjacent poles on the stator have opposite polarity and are interleaved with air gaps between them. The rotor is magnetised with six pairs of poles. The pairs are alternately N and S. The movement has a double worm reduction gear, a push and release starter lever, and a hand set knob. The locking plate chiming work and rack striking work are based on those used in Smith

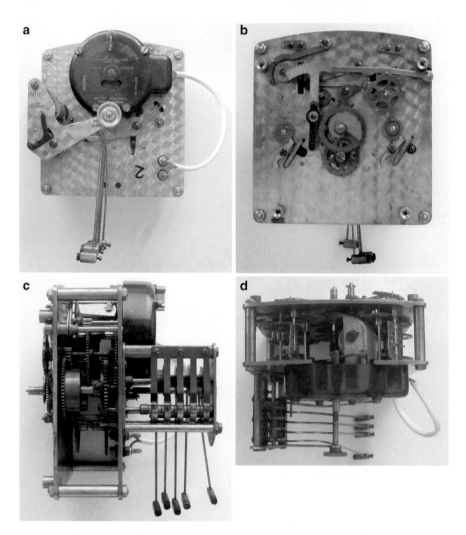

Fig. 11.42 Smith Type 1 chiming movement, (**a**) Rear view, (**b**) Front view, (**c**) Side view, (**d**) Top view

mechanical clocks (Smith 2008), except that the two mainsprings are automatically rewound by the synchronous motor rather than being hand wound. Chiming and striking is on gongs. Also see Sect. 3.4.

A sketch of the Type 1 Chiming movement is reproduced by Smith (2008) who states that is included in the 1933 catalogue. It is used in a clock with a date range of 1935–1939. Patent information on Type 1 chiming movements is: '366710, 369438, 374713, other patents pending'. Serial numbers 3797 and 3993 are known.

A rear view of the movement is shown in Fig. 11.42a. Moulded on markings on the motor cover include 'TRADE SEC MARK' and 'MADE IN ENGLAND'.

The serial number '3883' is on the lever below the motor. Figure 11.42b shows a rear view. The locking plate chiming work and rack striking work are visible. A side view is shown in Fig. 11.42c. The barrel for one of the mainsprings is visible. Fig. 11.42d shows a top view.

11.13.12 Smith Model 259 Chiming Movement

The Model 259 Chiming movement has a magnetised motor. There are 30 poles on the stator so the motor rotates at 200 rpm. Adjacent poles on the stator have opposite polarity and are interleaved with air gaps between them. The rotor is magnetised with six pairs of poles. The pairs are alternately N and S. The movement has a double worm reduction gear, a push and release starter lever, and a hand set knob. The rack chiming work and rack striking work are driven via a jockey wheel which engages when chiming and striking takes place. Chiming and striking is on gongs. Power is supplied via a 2 pin 2 A connector.

The Model 259 chiming movement was introduced in 1936 and is described by Miles (2011). British Patent information on Model 259 chiming movements is: '366710, 387108'.

A rear view of the movement is shown in Fig. 11.43a. Markings on the two parts of the back plate include 'SMITHS ENGLISH CLOCKS LTD CRICKLEWOOD LONDON', 'SEC REGD TRADE MARK' and '259' which is the movement Model number. Figure 11.43b shows a front view. The rack chiming work and rack striking work are visible. A view from underneath is shown in Fig. 11.43c. The underslung hammers mean that the gongs can be mounted horizontally. Figure 11.43d shows the gongs.

11.13.13 Smith Westminster Chiming Movement

The Westminster Chiming movement has a magnetised motor. There are 30 poles on the stator so the motor rotates at 200 rpm. Adjacent poles on the stator have opposite polarity and are interleaved with air gaps between them. The rotor is magnetised with six pairs of poles. The pairs are alternately N and S. The movement is self starting, but there is no hand set knob. The train splits with a double worm reduction gear to the motion work for the hands, and conventional wheels for the locking plate chiming work and the rack striking work. Chiming and striking is on gongs. The locking plate chiming work and rack striking work are based on those used in Smith mechanical clocks (Smith 2008), except that they are driven via a jockey wheel which engages when chiming and striking takes place, rather than by mainsprings. The plates are 4 in. (101.6 mm) high × 5-1/4 in. (133.4 mm) wide. The motor cover is 2-3/8 in. (60.3 mm) diameter × 1 in. (25.4 mm) deep.

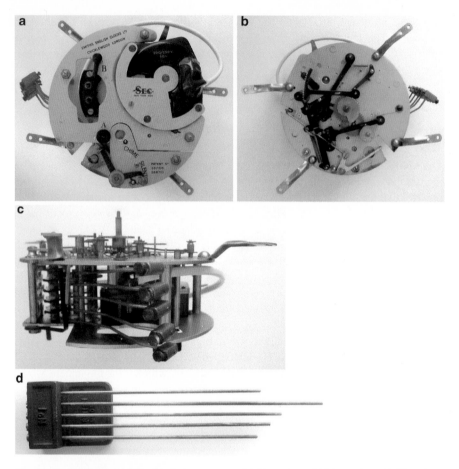

Fig. 11.43 Smith Model 259 chiming movement, (**a**) Rear view, (**b**) Front view, (**c**) View from underneath, (**d**) Gongs

The Westminster Chiming movement is described by Robinson (1940a) and in detail by Robinson (1940b). It was in current production in 1942 (Robinson 1942). It is used in a clock with a date range of 1953–1956. British Patent information on Westminster Chiming movements is: '366710, 387108, 412336, other patents pending'. A rear view of the movement is shown in Fig. 11.44a. The gong and hammer protectors are in position. These prevent damage while the clock is being transported, and are moved clear before the clock is put into use. Figure 11.44b shows a front view of the movement. The locking plate chiming work and rack striking work are visible. A top view of the movement is shown in Fig. 11.44c, d shows a view from underneath. The underslung hammers mean that the gongs can be mounted horizontally.

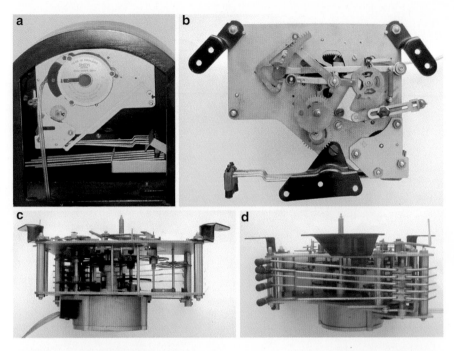

Fig. 11.44 Smith Westminster chiming movement, (**a**) Rear view, (**b**) Front view, (**c**) Top view, (**d**) View from underneath

11.14   Sterling Movement

The Sterling movement has a reluctance motor. The rotor has 30 poles so it rotates at 200 rpm. The reduction gear is similar to the going train of a conventional mechanical clock. The movement has a turn and release starter knob and a hand set knob at the back. The diameter of the plates is 2-5/8 in. (66.7 mm). There is no movement cover but the movement is enclosed by the clock case.

The Sterling movement is described by Robinson (1940a). Sterling clocks were advertised in 1937 (Anonymous 1937d) so the movement was in production at least from 1937 to 1940. Sterling clocks were exhibited at the 1947 British Industries Fair (Anonymous 1947), these featured a new Sterling movement. It is not clear whether this was a completely new movement or a revised version of the original movement. Either way it shows that Sterling movements were in production from 1937 to 1947.

A rear view of the movement is shown in Fig. 11.45a The back plate is marked '011353'. This is probably a serial number. Figure 11.45b shows a rear view with back plate removed. Figure 11.45c shows a front view, and Fig. 11.45d a top view.

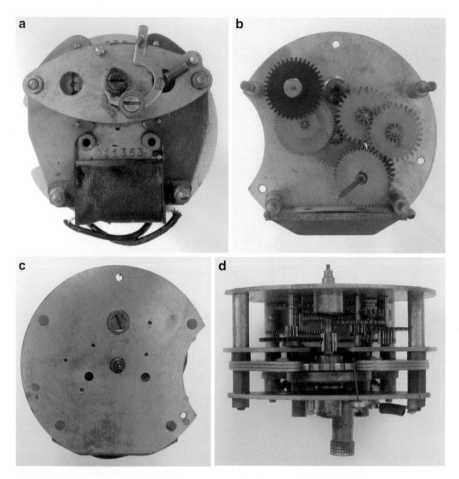

Fig. 11.45 Sterling movement, (**a**) Rear view, (**b**) Rear view with back plate removed, (**c**) Front view, (**d**) Top view

11.15   Temco Movements

Four types of Temco movements, made by the Telephone Manufacturing Company, are known. These are an early movement, for convenience this is referred to as the Mark I movement, and Marks II, IV and V movements. Marks I and II movements are for a 200–240 V supply and Marks IV and V for 200–250 V. Temco movements all have a cylindrical Bakelite movement cover, 3½ inches (88.9 mm) diameter and 2¼ inches (57.2 mm) deep. They are not self starting, and do not have provision for a seconds hand. With one known exception, the movements are not dated, so dates have to be based on circumstantial evidence. All Temco movements have 2 pin 2

Fig. 11.46 Temco Mark I
movement cover

A connectors. The female connector has 'TMC' moulded on. Most of the surviving Temco clocks have Mark V movements, others usually have Mark IV movements. Clocks fitted with Mark IV and Mark V movements are usually very reliable. Deluxe is an alternative name for the Mark V movement which was used in advertisements (Pook 2014).

11.15.1   Temco Mark I Movement

The Mark I movement has a reluctance motor. There are 20 poles on both stator and rotor so the motor rotates at 300 rpm. Unusually, all poles on the stator have the same polarity. The movement has a triple worm reduction gear, a press and release starter knob, and a push in hand set knob. Power is supplied via a 2 pin 2 A connector. The power consumption is ½ W (Philpott 1935).

It appears to be the early movement described briefly by Robinson (1942). A plausible for date its introduction is 1931 (Pook 2014).

Figure 11.46 shows a Temco Mark I movement cover. Moulded on markings include 'MADE IN ENGLAND BY THE TELEPHONE MFG. CO. LTD.', 'USE ONLY TMC ADAPTOR' and 'PATENTS APPLIED FOR'. The Telephone Manufacturing Co Ltd were awarded a patent in February 1932,

A front view of the movement is shown in Fig. 11.47a, and a rear view in Fig. 11.47b A rear view of the movement partly dismantled is shown in Fig. 11.47c and a three-fourth rear view in Fig. 11.47d.

Fig. 11.47 Temco Mark I movement (**a**) Front view, (**b**) Rear view, (**c**) Rear view, partly dismantled, (**d**) ¾ rear view, partly dismantled

11.15.2   Temco Mark II Movement

The Mark II movement has a reluctance motor. There are 20 poles on both stator and rotor so the motor rotates at 300 rpm. Unusually, all poles on the stator have the same polarity. The movement has a triple worm reduction gear, a press and release starter knob, and a push in hand set knob. Power is supplied via a 2 pin 2 A connector.

The Mark II movement is basically the same as the Mark I movement, but with some detail changes. A rear view of a Mark II movement cover is shown in Fig. 11.48a. Moulded on markings include 'MADE IN ENGLAND BY THE TELEPHONE MFG. CO. LTD.', 'TEMCO' 'USE ONLY TMC ADAPTOR',

Fig. 11.48 Temco Mark II movement, (**a**) Rear view of movement cover, (**b**) Rear view of movement

'PATENTS APPLIED FOR' and 'TYPE MARK II.' The serial number '15708' is hand written in white ink. The is the first movement on which the Temco trade mark is used. All known Mark II movements have serial numbers. A rear view of the movement is shown in Fig. 11.48b

11.15.3 Temco Mark IV Movement

The Mark IV movement has a magnetised motor. There are 30 poles on the stator so the motor rotates at 200 rpm. Adjacent poles on the stator have opposite polarity. They are interleaved with air gaps between them The rotor is magnetised with two pairs of poles. One pair is N and the other S. The movement has a double worm reduction gear, a starter knob which is rotated by hand, and a push in hand set knob. Power is supplied via a 2 pin 2 A connector. The diameter of the plates is 3¼ in. (82.6 mm). A bottom set version with a combined press and release starter and hand set knob at the bottom was available.

The magnetic design is described by Wise (1951). The same design is used for the Mark V movement.

The Mark IV movement is a successor to the Mark II movement in the sense that the dimensions of the movement cover are the same and the same type of connector is used. Hence, it is interchangeable with the Mark II movement. It is otherwise a completely new design. Production of the Mark IV movement probably started in 1933 (Pook 2014). It was still in production in 1938 when a brief description was published (Anonymous 1938a). Hence a plausible date range is 1933–1938 (Pook 2014).

Fig. 11.49 Rear view of a
Temco Mark IV movement
cover

A rear view of a Temco Mark IV movement cover is shown in Fig. 11.49. Moulded on markings include 'TEMCO', 'SET HANDS', 'TO START ROLL KNOB IN DIRECTION OF ARROW S', 'MADE IN ENGLAND BY THE TELEPHONE MFG. CO. LTD.' and 'TYPE MARK IV'. The serial number 32704 is handwritten in white near the top. Not all Mark IV movements have serial numbers. The knurled starter knob is mostly enclosed, its lowest part is just visible at the centre, below the moulded on arrow. The Bakelite hand set knob is on the left, and the two pin connector at the bottom. Mark IV movements were made with at least two different hand spindle lengths to suit different cases. Figure 11.50 illustrates some of the features of a Mark V movement. The figure equally well illustrates features of a Mark IV movement.

A 1937 advertisement (Anonymous 1937b) states that "Temco" clocks have high-torque, easy starting motors fitted with self-lubricating phosphor bronze bearings. The "Temco" worm-drive transmission is self-cleaning and efficient. "Temco" clocks are made in a factory specialising in electrical apparatus and precision instruments'. In 1938 the Mark IV movement was described as 'one of the best movements on the market' (Anonymous 1938b).

11.15.4   Temco Mark V Movement

The Mark V movement has a magnetised motor. There are 30 poles on the stator so the motor rotates at 200 rpm. Adjacent poles on the stator have opposite polarity. The rotor is magnetised with two pairs of poles. The movement has a double worm reduction gear. The back set version has a starter knob which is rotated by hand, and a push in hand set knob. Power is supplied via a 2 pin 2 A connector. The diameter of the plates is 3¼ in. (82.6 mm). Mark V movements were made with at least three different hand spindle lengths. A bottom set version with a combined press and release starter and hand set knob at the bottom was available. Also see Sect. 3.2.2.

The back set Mark V movement is similar to the Mark IV movement. It is not clear why it was given a different designation because variations within a particular mark are comparable to differences between the two marks. Brief descriptions are given by Robinson (1940a, 1942) and Wise (1951).

The differences between the Mark IV and back set Mark V movements are minor (Pook 2014). They appear to be confined to differences in pivot diameters, different methods of fixing the motor to the back plate, and different methods of fixing the hand set knob. Some pivot diameters also differ between versions of the Mark IV movement and between versions of the Mark V movement. Nevertheless, there is some interchangeabilty of parts, including movement covers, within and between Mark IV and Mark V movements.

A rear view of the Bakelite cover of a back set Mark V movement is shown in Fig. 11.50a. Except for 'TYPE MARK V', moulded on markings on Mark V movement covers are the same as for Type IV. Serial numbers never appear on Mark V movement covers. The hand set knob on the left is not original. A front view of the movement is shown in Fig. 11.50b. The tell tale dial is visible. Figure 11.50c shows a front view with the front plate and some wheels removed. The congealed oil visible on the fibre wheel is typical of synchronous clocks that have stopped working. A side view of the movement is shown in Fig. 11.50d The motor is almost entirely enclosed by its cover. The two pin connector is at the bottom, and the knurled starter knob is on the right. A view from underneath is shown in Fig. 11.50e. The spring which provides the friction drive to the hands is near the centre. Figure 11.50f shows a rear view, and Fig. 11.50g a rear view with the back of the motor cover removed. The rotor and the interleaved poles on the stator are visible. The coil is inside the poles. The back part of the cover is a friction fit, so it is easily removed and replaced. The cover keeps the movement clean (Wise 1951), which must contribute to the longevity of Mark V (and Mark IV) movements.

Temco movements are not normally dated, but Fig. 11.50h shows a detail of the front of a back set Mark V movement. This is ink stamped '21 JUL 1938' on the left. What appears to be a circular inspector's mark on the right is too faint to be deciphered. This suggests that the changeover from the Mark IV movement to the Mark V movement took place during 1938. A clock of a style fitted with a back set Mark V movement was shown at the British Industries Fair in May 1948 (Anonymous 1948). A plausible date range for the Mark V movement is 1938 to the late 1940s (Pook 2014).

A rear view of the Bakelite cover of a bottom set Mark V movement is shown in Fig. 11.51a. Moulded on markings are the same as those on a back set Mark V movement. Starting and hand setting is by a knob at the end of the shaft below

the bottom. A rear view of the movement is shown in Fig. 11.51b. A front view
with the front plate and some wheels removed is shown in Fig. 11.51c. Changes
have been made to accommodate different hand setting and starting arrangements
(cf Fig. 11.50c).

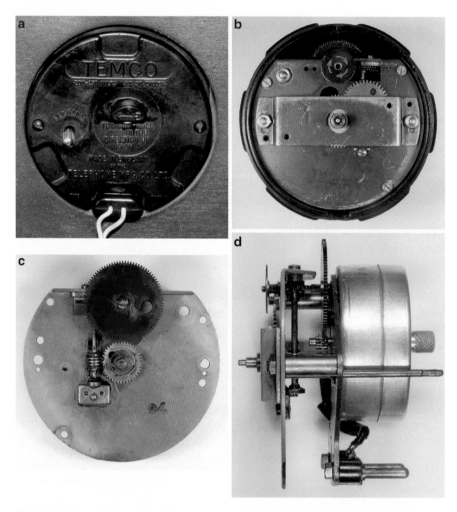

Fig. 11.50 Temco back set Mark V movement, (**a**) Rear view of movement cover, (**b**) Front view,
(**c**) Front view with the front plate and some wheels removed, (**d**) Side view, (**e**) View from
underneath, Rear view, (**f**) Rear view, (**g**) Rear view with back of the motor cover removed, (**h**)
Detail of front view showing ink stampings

Fig. 11.50 (continued)

11.16 Westclox Type No. BM25 Movement

The Type No. BM25 movement has a magnetised motor. There are 30 poles on the rotor so it rotates at 200 rpm. The rotor consists of two discs, each with 15 poles bent over so that they are interleaved. They are magnetised by a disc magnet between them. Hence the poles are alternately N and S. The reduction gear is similar to the going train of a conventional mechanical clock. The movement is self starting with a hand set knob on the bottom. The movement cover is 3-5/8 in. (92.0 mm) diameter × 1-1/8 in. (28.6 mm) deep plus an integral hanging loop. A back set version was also made.

The BM25 movement is not included in the list of Westclox synchronous movements in ClockDoc (2014). The movement cover for the bottom set version of the Type No. BM25 movement is shown in Fig. 11.52a and that for the back set version in Fig. 11.52b. Moulded on markings on both versions include 'TYPE NO. BM25' and 'MADE IN SCOTLAND'.

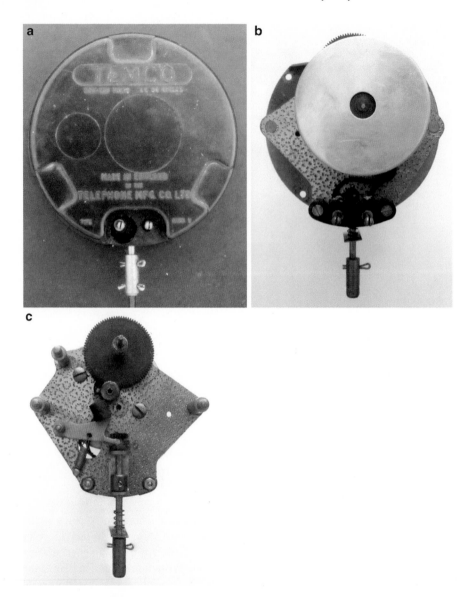

Fig. 11.51 Temco bottom set Mark V movement, (**a**) Rear view of movement cover, (**b**) Rear view, (**c**) Front view with the front plate and some wheels removed

A front view of a Type No. BM25 bottom set movement is shown in Fig. 11.53a. The front plate is marked 'U 8 67'. '8 67' is probably the date of manufacture, August 1967. Figure 11.53b shows a front view with the front plate removed. The interleaved poles on the rotor are visible. A rear view of the back plate is shown in Fig. 11.53c.

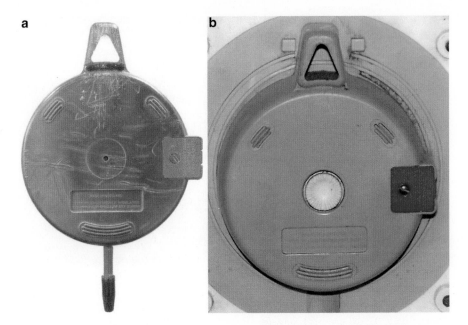

Fig. 11.52   Westclox Type No. BM25 movement covers, (**a**) Bottom set, (**b**) Back set

11.17   Conclusions

1. Most manufacturers made only a small number of different synchronous movements, with some variations to suit different clocks. The exception is Smith who
   made at least 13 different movements, together with a range of variations to suit
   different clocks.
2. Both reluctance motors and magnetised motors can be satisfactory in service.
   Small manufacturers used reluctance motors. Magnetised motors were only used
   by large manufacturers, presumably because the development costs could be
   amortised over a large number of units.
3. The rotation speed of motors was a compromise. The most popular was 200 rpm,
   but this was not universal.
4. Synchronous motors are not inherently self starting. Many clocks had manual
   start motors. Some had self starting motors, but reliability of self starting methods
   was sometimes a problem.
5. Two types of reduction gear were used. In one type the reduction gear is similarly
   to the going train of a conventional mechanical clock. In the other type one, two
   or three worm gears were used. Both types can be satisfactory in service.

Fig. 11.53 Westclox Type No. BM25 bottom set movement, (**a**) Front view, (**b**) Front view with front plate removed, (**c**) Rear view of back plate

References

Anonymous (1932) BS 472: 1932 Mains-operated synchronous clocks. British Standards Institution, London
Anonymous (1935) Smith's clocks. Horological J 78(926):11
Anonymous (1936) Smith clocks. Horological J 78(934):9
Anonymous (1937a) Synchronous clock unit with Geneva stopwork. Horological J 79(948):12–14
Anonymous (1937b) Temco synchronous electric clocks. Horological J 79(947):7
Anonymous (1937c) Smith's English clocks. Striking developments. Horological J 79(942):28

Anonymous (1937d) Sterling Clock Co. Ltd. Horological J 80(949):11

Anonymous (1938a) "Temco" Mark IV synchronous clock movement. Horological J 80(955): 14–15

Anonymous (1938b) Better, cheaper, synchronous clocks. Horological J 81(961):38–39

Anonymous (1939) Britain's newest clock factory. Horological J 81(974):397–422

Anonymous (1947) Horological section of the B.I.F. A brief summary of the main exhibits. Horological J 89(1064):235–237

Anonymous (1948) Review of the 1948 British industries fair. Horological section. Horological J 90(1076):280–283

Anonymous (1962) BS 472: 1962 Mains-operated synchronous clocks. British Standards Institution, London

Ball AE (1932) Electric clocks which operate on service mains. Chapter X. Horological J 74(883):109–111, 135

Bird C (2003) Metamec. The clockmaker. Dereham. Antiquarian Horological Society, Ticehurst

British Patent 125953 (1920) Improvements in and relating to electric time indicating apparatus

British Patent 366710 (1932) Improvements relating to electric motors

British Patent 369438 (1932) Improvements relating to electric clocks

British Patent 374713 (1932) Improvements relating to electric clocks

British Patent 377696 (1932) Improvements in alternating current motors

British Patent 379383 (1932) Improvements in and relating to synchronous electric motors

British Patent 384441 (1932) Improvements relating to electric clocks

British Patent 384681 (1932) Improvements in and relating to synchronous electric motors

British Patent 384880 (1932) Improvements in electric motors

British Patent 386756 (1933) Synchronous electric motors and means for starting them

British Patent 387108 (1933) Electric clocks

British Patent 412336 (1934) Improvements relating to synchronous electric motors

British Patent 444688 (1936) Improvements in and relating to synchronous-motor electric clocks

British Patent 484222 (1938) Improvements relating to small synchronous electric motors

British Patent 571849 (1945) Improvements in or relating to uni-directional drive mechanism

British Patent 744204 (1956) Improvements in or relating to the control of the rotational direction of synchronous electric motors

British Patent 806383 (1958) Improvements in methods of joining parts

Britten FJ (1978) The watch & clock makers' handbook, dictionary and guide. 16th edn. Revised by Good R. Arco Publishing Company, New York

ClockDoc – the electric clock archive. http://www.electricclockarchive.org/. Accessed 2014

Eves J (1944) Starting trouble. Horological J 86(1032):295

Lines MA (2012) Ferranti synchronous electric clocks. Paperback edition with corrections. Zazzo Media, Milton Keynes

Miles R (2003) The movements of the electric clocks marketed by Metamec. In: Bird C (ed) Metamec. The clockmaker. Dereham. Antiquarian Horological Society, Ticehurst, pp 87–112

Miles RHA (2011) Synchronome. Masters of electrical time keeping. The Antiquarian Horological Society, Ticehurst

Philpott SF (1935) Modern electric clocks, 2nd edn. Sir Isaac Pitman & Sons Ltd, London

Pook LP (2014) Temco Art Deco domestic synchronous clocks. Watch Clock Bull 56/1(407): 47–58

Robinson TR (1940a) Servicing the synchronous clock. Horological J 82(984):13–36

Robinson TR (1940b) New Smith "Sectric" chime clock. Horological J 82(980):172–175

Robinson TR (1942) Modern clocks. Their repair and maintenance, 2nd edn. N A G Press Ltd, London

Robinson TR (1946) The Ferranti rotor modified. Horological J 88(1056):372

Robinson TR (1954) The Ferranti synchronous timepiece. Horological J 96(1153):641–642

Robinson TR (1956a) The Metamec synchronous clock. Horological J 98(1174):417–418
Robinson TR (1956b) Keeping up with clock improvements. Horological J 98(1171):232–234
Smith B (2008) Smiths domestic clocks, 2nd edn. Pierhead Publications Limited, Herne Bay
Wemyss HD (1933) The tick in electric clocks. Horological J 79(902):401
Wise SJ (1951) Electric clocks, 2nd edn. Heywood & Company Ltd., London